The Lion that
DIDN'T ROAR

CAN THE KIMBERLEY PROCESS STOP
THE BLOOD DIAMONDS TRADE?

The Lion that DIDN'T ROAR

CAN THE KIMBERLEY PROCESS STOP THE BLOOD DIAMONDS TRADE?

NIGEL DAVIDSON

Australian
National
University

PRESS

ANU PRESS

Published by ANU Press
The Australian National University
Acton ACT 2601, Australia
Email: anupress@anu.edu.au
This title is also available online at press.anu.edu.au

National Library of Australia Cataloguing-in-Publication entry

Creator:	Davidson, Nigel, author.
Title:	The lion that didn't roar : can the Kimberley Process stop the blood diamonds trade? / Nigel Davidson.
ISBN:	9781760460259 (paperback) 9781760460266 (ebook)
Subjects:	Conflict diamonds--Africa.
	Diamond mines and mining--Africa.
	Diamond industry and trade--Africa.
	Human rights--Africa.
Dewey Number:	338.88722382

Cover design and layout by ANU Press. Cover photograph: 'Child in a rebel camp in the north-eastern Central African Republic', photographer Pierre Holtz (by hdptcar www.flickr.com/photos/hdptcar/949798984/).

Contents

Preface vii

Acknowledgements xi

Acronyms xiii

Glossary: Diamond Industry Terms xvii

1.   Introduction: Showdown at Kinshasa 1

2.   Are Conflict Diamonds Forever?: Background to the Problem . . .21

3.   The Kimberley Process: Did the Lion Roar? 73

4.   Kimberley at the National Level: Fancy Footwork? 127

5.   Growing Teeth: The UN Security Council and
     International Tribunals 147

6.   Raging Bulls and Flyswatters: The Networked
     Pyramid Model 173

7.   The Dual Networked Pyramid Model: The Pyramid
     Inside the Pyramid 213

8.   Applying the Dual Networked Pyramid Model:
     Naming, Shaming, and Faming 225

9.   Did You Hear Something?: Concluding Remarks 255

Appendix: The Kimberley Process Core Document 265

Bibliography 283

Preface

The year 2017 will be the one that Australia assumes the duty of being Chair of the Kimberley Process Certification Scheme (KP). The KP is an international organisation that regulates the world's diamond trade. Diamonds are a symbol of love, purchased to celebrate marriage in many parts of the world, but this trade has been linked with warfare and human rights violations committed in African producer countries such as Sierra Leone, Côte d'Ivoire and Angola. Graphic accounts of murder and mayhem, fuelled by the diamond black market, continue to emerge from the Democratic Republic of Congo, Zimbabwe, and the Central African Republic, posing an existential threat to the multibillion dollar industry. These human rights violations fall under the legal categories of war crimes, crimes against humanity, and genocide, the most serious crimes under international law. In response to the grim reality of the blood diamonds trade, De Beers and other major corporate players joined with non-governmental organisations and national governments to create the Kimberley Process Certification Scheme in 2002. The objective of the Kimberley Process is to distinguish the legitimate rough diamond trade from the trade in diamonds linked to serious human rights abuses, known as conflict diamonds or blood diamonds. The Kimberley Process involves a system of export and import certificates attesting to the clean character of rough diamonds, and is backed up by a peer review system to monitor compliance. The Kimberley Process has been supported through the regulatory action of national governments at the domestic level, as well as the United Nations Security Council (UNSC) and the International Criminal Court (ICC) internationally.

The first research question considered by this book is: to what extent has the conflict diamonds governance system achieved its objectives? In response, it can be said that the conflict diamonds governance system has made significant progress in its core mandate. The quantity and value of the international legitimate diamond industry, once the very paradigm of secrecy, has become more transparent through publicly available Kimberley Process statistics. Based on these statistics, the Kimberley Process estimates that the blood diamond trade now constitutes less than 1 per cent of the world's rough diamond trade. However, it has not always been smooth sailing for the Kimberley Process, which has recently arrived in particularly stormy waters. The integrity of the system has been endangered by the seeming inability of the Kimberley Process to take appropriate action in the face of serious non-compliance by three important national government stakeholders: Venezuela, Zimbabwe, and Angola. Commentators are asking whether the Kimberley Process lion has forgotten how to roar.

The second research question is: does an application of the networked pyramid regulatory model to the system provide descriptive or normative insights into its effectiveness? In considering the relative success, and the current challenges facing the conflict diamonds governance system, important insights may be gained by looking at the system with reference to the networked pyramid regulatory model. Before applying the model, the book suggests a modification, dubbed the dual networked pyramid model (DNPM), whereby the micro-regulatory system at the national level is seen as a networked pyramid within the greater networked pyramid of the international system. The relative success of the Kimberley Process to date, when analysed against this theoretical hybrid of network and pyramid models, is largely linked to its self-conscious incorporation of insights from networks theory. At the international level, the Kimberley Process can be seen as the central node, or command centre, in which information is gathered, and regulatory action coordinated, from networks of corporations, national governments, and non-governmental organisations. Its relative success to date can largely be attributed to a process of socialisation whereby big business and most national governments have become key supporters.

It is, however, in the theoretical domain of the regulatory pyramid that the Kimberley Process might find a way out of its current deadlock. Pyramid theory recognises the primacy of soft power, such as dialogue and socialisation, but demands escalation to more coercive measures where regulated parties are unresponsive or recalcitrant. It is suggested that improved procedures for managing serious non-compliance, combined with an agreed pathway to expulsion from the Kimberley

Process in such cases, would bring the Kimberley Process into better alignment with the pyramid model and help it to move out of the log jam in which it currently finds itself. A more defined pathway of escalation to the UNSC and the ICC would bolster the ongoing efficacy of the conflict diamonds governance system. A recommended mechanism for doing this would be to amend the Statute of the ICC to include a crime of trafficking in conflict diamonds, to be defined in terms of contravening a UNSC diamond embargo.

Beyond breaking the current deadlock, the Kimberley Process has an opportunity to reinvent itself by embracing the concept of development diamonds. First suggested by non-governmental organisations, this label might be applied to diamonds from the informal sector that are not merely free from the taint of international crime, but also comply with other human rights standards, most notably freedom from child labour. A further modification to the DNPM, assisted by insights from the pyramid of rewards theoretical model, reveals that the Kimberley Process has the chance to systematically ratchet up human rights, and health, safety and environmental standards in the artisanal sector, thereby buttressing the industry against the return of blood diamonds.

This book is a revised version of a thesis submitted towards the degree of Doctor of Juridical Science at The Australian National University. I was awarded the degree in July 2012, but the material in the manuscript has been updated to reflect developments as at the end of December 2015.

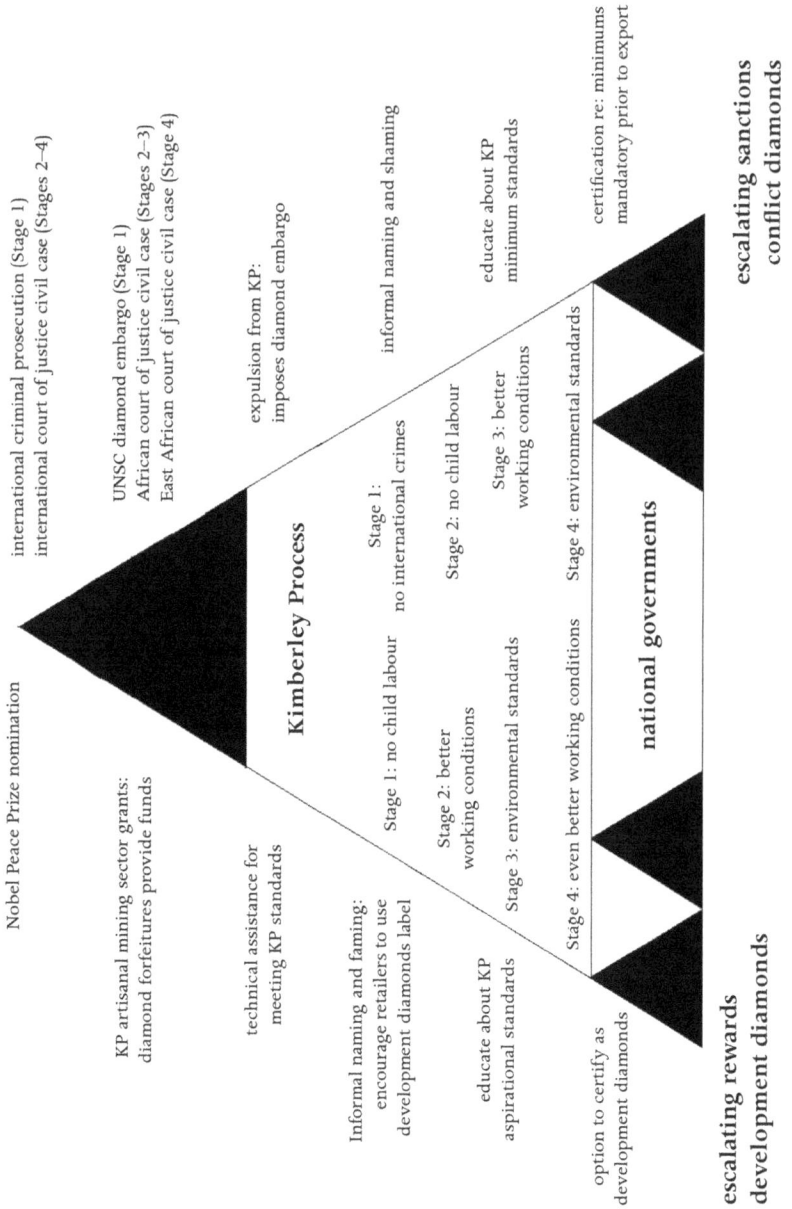

Figure 1: The Dual Networked Pyramid Model as Applied to the Conflict Diamonds Governance System

Source: Author's research.

Acknowledgements

Many thanks to my father, Dr John Davidson, and my sister, Rosalind Ta'eed, for their support in this endeavour. This work was initially prepared as a doctoral thesis with The Australian National University. I would like to acknowledge the significant contributions of my supervisors, Professor Andrew Byrnes and Professor Peta Spender, as well as those of Dr Gregor Urbas and Professor John Braithwaite.

It is my hope that this book will contribute to the global debate about the blood diamonds problem, and help to stimulate renewed interest in safeguarding and improving the lives of the many people affected. To do so, I believe we must look beyond economic value to the intrinsic value of human life and dignity, which is synonymous with a spiritual perspective:

> It is said that in South Africa, a diamond mine is discovered. Although this mine is most valuable, yet after all it is stone. Perchance, God willing, the mine of humanity may be discovered and the brilliant pearls of the Kingdom be found.

> Abdu'l-Baha (1844–1921)

Acronyms

ADFL	Alliance of Democratic Forces for the Liberation of Congo (militia that became the Government of the Democratic Republic of Congo)
AFRC	Armed Forces Revolutionary Council (Sierra Leone, rebel militia)
ASIC	Australian Securities and Investments Commission
CAR	Central African Republic
CDF	Civilian Defence Forces (Sierra Leone, pro-government militia)
CFCs	Chlorofluorocarbons
DDII	Diamond Development Initiative International
DNPM	Dual networked pyramid model
DRC	Democratic Republic of Congo
DRI	Directorate of Revenue Intelligence (India)
ECOMOG	Economic Community of West African States Cease-Fire Monitoring Group
EITI	Extractive Industries Transparency Initiative
FNI	*Front des Nationalistes et Intégrationnistes* (Front for National Integration) (Congo, rebel militia)
FNLA	National Liberation Front of Angola (rebel militia)
FRPI	*Force de Résistance Patriotique en Ituri* (Patriotic Force of Resistance in Ituri) (Congo, rebel militia)
GATT	General Agreement on Tariffs and Trade

ICC	International Criminal Court
IFAC-3	Industry Functional Advisory Committee on Intellectual Property Rights for Trade Policy Matters (US)
ILO	International Labour Organization
IMF	International Monetary Fund
KP	Kimberley Process Certification Scheme, also known as the Kimberley Process
MLC	Congolese Liberation Movement
MONUC	United Nations Mission in the Democratic Republic of Congo
MPLA	Popular Movement for the Liberation of Angola (militia which became the Government of Angola)
MRC	*Mouvement Révolutionnaire du Congo*
NGO	Non-governmental organisation
OCHA	United Nations Office for the Coordination of Humanitarian Affairs
OECD	Organisation for Economic Co-operation and Development
RCB	Republic of Congo-Brazzaville
RCD	Congolese Rally for Democracy (Congo, rebel militia, which later split into three groups: RCD-Goma, RCD-National, and RCD-ML)
RUF	Revolutionary United Front (Sierra Leone, rebel militia)
UN	United Nations
UNDP	United Nations Development Programme
UNGA	United Nations General Assembly
UNICEF	United Nations Children's Fund
UNITA	National Union for the Total Independence of Angola (rebel militia)

UNOCI	United Nations Operation in Côte d'Ivoire
UNOMSIL	United Nations Observer Mission Sierra Leone
UNSC	United Nations Security Council
UPC	*Union des Patriotes Congolais* (Union of Congolese Patriots) (Congo, rebel militia)
USAID	United States Agency for International Development
WDC	World Diamond Council
WTO	World Trade Organization

Glossary: Diamond Industry Terms[1]

Alluvial	The name of a type of diamond and the type of shallow mine it is extracted from, with diamonds found in river beds and shallow deposits. A form of mine that can be exploited by artisanal techniques.
Artisanal miner	A small-scale miner of alluvial deposits, whose tools are typically simple, such as shovels and hand sieves.
The four Cs	Colour, clarity, cut, and carat. These four factors are considered when valuing a stone.
Carat	Unit of measurement of a diamond. There are five carats to one gram. Diamonds vary from a fraction of one carat up to a very rare couple of thousand. In trade, a stone of 60 carats would be considered large.
Colour	Diamonds come in a wide range of hues, tints, and colours. They can be described as whitish, yellowish, greenish, brownish, pinkish, bluish, and so on. Stones from different countries can vary in colour.
Comptoir (Angola)	Small-scale diamond buyers who act as middlemen.

1   Based on Global Witness, 'A Rough Trade: The Role of Companies and Governments in the Angolan Conflict' (Report, Global Witness Ltd, 1998) 15.

Garimpeiros (Angola)	Illegal miners, usually artisanal.
Gem quality	The highest quality of diamond, which is normally in high demand and commands top prices.
Inclusions	The particles and matter sometimes found within a diamond.
Kimberlite	The name of a type of diamond and the type of mine it is extracted from, which has deep subterranean volcanic pipes.
Mixed parcel	A parcel of rough diamonds from more than one country.
Parcel	A quantity of diamonds, which can vary from 10 carats up to thousands of carats.
Polished	The term used to describe stones when they have been worked. Up to 50 per cent of the diamond can be lost when polished, depending on the shape of the stone.
Rough	Rough diamonds are unworked and appear in their natural state.

1

Introduction: Showdown at Kinshasa

The shallow tunnel where my colleagues were working collapsed and trapped them inside. There was nothing I could do to save them; I had to run for my own life. On that night, three people were shot by police and died on the field. The following morning, police ordered us to bury the three bodies in one of the pits on the field. When I asked to dig out my four colleagues, a police officer told me, 'Consider them already buried'.

Artisanal miner in Marange, Zimbabwe[1]

On 23 June 2011, at Kinshasa in the Democratic Republic of Congo (DRC), civil society delegates staged a dramatic walkout from a meeting of the Kimberley Process (KP). At the meeting, the KP chairperson had acted to endorse the sale of controversial Zimbabwean blood diamonds despite a lack of consensus by KP participants. In a statement issued after the walkout, the civil society coalition, representing a range of non-governmental organisations (NGOs), stated:

---

1   Human Rights Watch interview with Marange artisanal miner R M, Harare, 26 February 2009, referring to an attack in August 2008, cited in Human Rights Watch, 'Diamonds in the Rough: Human Rights Abuses in the Marange Diamond Fields of Zimbabwe' (Report, June 2009).

> The agreement between the Kimberley Process and Zimbabwe being discussed this week falls far short of what is acceptable to maintain the credibility of the Kimberley Process, protect civilians and civil society members living and working in Marange or prevent substantive quantities of illicit diamonds from infecting the global diamond supply chain.[2]

To better understand why the NGOs walked out of the meeting on that day, it is necessary to understand the nature of the Kimberley Process, an organisation that was established to tackle the issue of conflict diamonds, also known as blood diamonds. Blood diamonds constitute a segment of the rough diamond trade that is linked to egregious human rights violations in a number of African diamond mining countries.[3] These diamonds are known as blood diamonds because of their connection with groups that have used enforced labour, recruited and deployed children as soldiers, murdered and raped civilians, amputated the limbs of their victims, and terrorised civilian populations, often as part of waging civil war. Although diamond-fuelled violence has diminished with the emergence of peace in Angola, Sierra Leone, and Côte d'Ivoire, there are ongoing concerns relating to the war-torn DRC and the Central African Republic.

What has garnered recent international attention more than any other blood diamond issue, however, is the violence associated with the discovery of diamonds in the Marange region of Zimbabwe. With civilian casualties in the hundreds, the brutality of the management of the Marange diamond fields by Zimbabwe's police and armed forces has become well known to the international community. What would appear to be a clear violation of the KP's mandate, which is to prevent such blood diamonds being traded on the international market, instead attracted a different response from the KP. Rather than excluding these diamonds from the international market and expelling Zimbabwe from the KP, the KP chair controversially acted to mandate the sale of several shipments of these diamonds, despite a lack of consensus within the

---

2   Partnership Africa Canada, 'NGOs Walkout of Kinshasa KP Meeting, Consider Options', *Other Facets: News and Views on the International Effort to End Conflict Diamonds* (No. 35, August 2011), 1. Available at: www.pacweb.org/images/PUBLICATIONS/Other_Facets/OF35-eng.pdf; Gooch, C, 'Global Witness Founding Director's Statement on NGO Coalition Walk-Out from Kimberley Process Meeting', 27 June 2011. Available at: www.globalwitness.org/en/archive/global-witness-founding-directors-statement-ngo-coalition-walk-out-kimberley-process/.

3   It should be noted that the definition of a 'conflict diamond' or 'blood diamond' was itself part of the dispute in the Zimbabwe Marange diamonds dispute. This is discussed in Chapter 2.

organisation. Concerned that the core mandate of the KP was being contravened, the NGOs stormed out of the Kinshasa meeting, although they have said that they will remain within the organisation, at least for the moment.[4] To put it metaphorically, NGOs were wondering whether the KP lion had lost its roar.

The so-called showdown at Kinshasa provides a useful point of reference in seeking to analyse and assess the relative strengths and weaknesses of the Kimberley Process at this moment of institutional crisis. The challenges facing the Kimberley Process are particularly relevant to an Australian audience, as Australia will take its turn as Chair of the KP in 2017. The Kimberley Process Certification Scheme (KP) was established in 2002 as the international community's primary response to the blood diamond problem. It has mobilised the energies of civil society, the major corporate players in the rough diamond trade, and national governments. The Kimberley Process is a chain of custody arrangement, which aims to provide a warranty as to the origin of each diamond from the point of mining, through to export and polishing, to incorporation in jewellery and final sale to a consumer in a retail context. The primary mechanism for this guarantee about the origin of the diamonds is the export certificate, which guarantees that a package of rough diamonds is conflict-free when it leaves the original producer country. Compliance with Kimberley requirements is monitored through review visits by delegations, involving representatives of civil society, industry, and government. In cases of serious non-compliance, the Kimberley Process has the ability to suspend or expel a government member, meaning that they are excluded from the legitimate rough diamond trade.

The United Nations Security Council (UNSC) and the international criminal tribunals are the other major players in the conflict diamonds governance system. The UNSC has played an important monitoring role through its expert committee reports, and has imposed legally binding sanctions on diamonds from problem countries in a number of instances. It is arguable that the UNSC was the midwife of the Kimberley

---

4   Partnership Africa Canada, 'NGOs Walkout of Kinshasa KP Meeting, Consider Options', *Other Facets: News and Views on the International Effort to End Conflict Diamonds* (No. 35, August 2011), 1. Available at: www.pacweb.org/images/PUBLICATIONS/Other_Facets/OF35-eng.pdf; Gooch, C, 'Global Witness Founding Director's Statement on NGO Coalition Walk-Out from Kimberley Process Meeting', 27 June 2011. Available at: www.globalwitness.org/en/archive/global-witness-founding-directors-statement-ngo-coalition-walk-out-kimberley-process/.

Process, facilitating its birth. The international criminal tribunals, namely the Special Court for Sierra Leone and the International Criminal Court (ICC), provide a further level of conflict diamonds governance. A growing number of international prosecutions by these bodies, most notably the Charles Taylor case, have shone a spotlight on the role played by conflict diamonds in the perpetration of crimes against humanity and war crimes.

Research Questions and Main Argument

The walkout at Kinshasa highlights the recent state of crisis that the KP finds itself in, almost 13 years after its creation in November 2002. Since the creation of the KP, a significant body of articles and reports has been written by academics and NGO activists. This has largely focused on the blood diamonds problem, the wider context of the resources and conflict linkage, and practical evaluations of the KP. Only two monographs have been published about the KP to date, the insider/practitioner perspective of Ian Smillie in his work *Blood on the Stone,* and the doctoral dissertation in book format of Franziska Bieri, entitled *From Blood Diamonds to the Kimberley Process: How NGOs Cleaned up the Global Diamond Industry.* Bieri's work involved the collation of interview material, and focuses on the role of NGOs in identifying the issue of blood diamonds, campaigning on the issue and ultimately providing a guiding role in the creation of the tripartite KP.

My book represents an original contribution to the field, as distinct from the works of Smillie and Bieri. Smillie presented a practitioner/ insider perspective on the Kimberley Process, while Bieri's academic work focused on the role of NGOs in relation to the Kimberley Process. By contrast, my work analyses the KP in the context of other international regulatory mechanisms, using an original theoretical model (the dual networked pyramid model (DNPM)), which assists in understanding its successes thus far, as well as suggesting ways in which the system might be improved upon. Other works discuss the blood diamonds problem in general, without throwing specialist light on the Kimberley Process. In this category is Greg Campbell's work *Blood Diamonds: Tracing the Deadly Path of the World's Most*

Precious Stones.[5] By contrast with Smillie and Bieri's specialist work on the Kimberley Process, and the generalist blood diamonds literature, my book gives a rigorous overview of both the conflict diamonds and regulatory theory literature, provides systematic analysis of that literature, and provides theoretical modelling.

The first research question that this book posits is: to what extent has the conflict diamonds governance system achieved its objectives? In considering this question, the work discusses not only the Kimberley Process but also other international institutions that have played a central role in conflict diamonds governance — in particular the UNSC and the ICC (including its sister tribunals). It seeks to be, arguably, the first large-scale work to not only describe the role that each of these bodies has played in relation to conflict diamonds governance, but the way in which they have interacted, and how these interactions could be improved.

In considering the interactions between the major institutional players in the conflict diamonds governance system, the book seeks to be the first work to apply a theoretical, regulatory model to that system, with a view to understanding it better. As such, the book involves consideration of a further research question: does an application of the networked pyramid regulatory model to the conflict diamonds governance system provide descriptive or normative insights into its effectiveness? The networked pyramid model suggests that the most successful regulatory approaches extend beyond governmental action alone, to embrace non-governmental actors such as civil society organisations and business entities. It argues that the most significant regulatory gains are made through the horizontal techniques of dialogue, persuasion, and socialisation. Nevertheless, the model recognises that the deployment of vertical coercive interventions may be necessary in appropriate circumstances.

Despite the intuitive applicability of the networked pyramid model, it is arguable that modifications to the model may be desirable before it can usefully be applied to the conflict diamonds governance system. The book suggests two significant modifications to the model that result in a so-called dual networked pyramid model. It is dual in two senses: firstly, it incorporates regulatory systems at both the national

---

5   Campbell, G, *Blood Diamonds: Tracing the Deadly Path of the World's Most Precious Stones* (Basic Books, 2013).

and international levels, creating a pyramid inside the pyramid that models national governments as both regulators and the subjects of regulation. Secondly, by incorporating insights from the pyramid of rewards, it models both rewards and sanctions in a single model. For these reasons, it can be argued that the dual model is well placed to generate theoretical insights into a range of complex international systems, such as the global intellectual property standards system, beyond the application that is the subject of this book.

In relating this model to the conflict diamonds system, the second research question considers whether the DNPM provides descriptive or normative insights into the operation of the conflict diamonds governance system. In particular, it explores whether the reasons for the successes or failures of the system can be linked to the way in which it incorporates and implements features of the DNPM.

Returning to the first research question, the book, which considers developments up to the end of December 2015, argues that a significant degree of achievement can be attributed to the conflict diamonds governance system. It is argued that the governance system has contributed to the diminution of the conflict diamonds trade from estimates as high as 15 per cent in the 1990s, when the Sierra Leone and Angolan conflicts were active, to less than 1 per cent of the world's rough diamond trade in recent years. When considering the role of the Kimberley Process, this relative achievement is due in large measure to its ability to enlist the support of the major diamond mining players, the vast majority of national governments involved in rough diamond production, trading and polishing, as well as committed international NGOs. KP has not been an unqualified success, however. The fact that Angolan and Zimbabwean conflict diamonds have not been effectively filtered out of the legitimate trading system, as well as the opening of a smuggling gateway created by Venezuela's active opposition, show that serious challenges remain for the conflict diamonds governance system. These problems have arisen through the inability of the KP to expel member governments in situations of serious non-compliance. It is arguable that they can only be effectively addressed through appropriate application of vertical coercive interventions, as suggested by the pyramid features of the networked pyramid regulatory model. As such, the book recommends that Kimberley Process procedures be reformed so as to enable timely and definitive expulsion of recalcitrant governments. The work recommends that a new, specific crime of trafficking in conflict diamonds be created in the provisions of the

Rome Statute of the ICC. Such a crime should be defined in terms of contravening a UNSC ban on diamond trading and would, it is argued, strengthen the conflict diamonds governance system at all levels.

Beyond resolving the initial crisis that the KP finds itself in, it would arguably benefit from a renewed mandate, focusing on the concept of development diamonds. The book explores these possibilities, with reference to not only regulatory pyramid insights, but also insights from the incentive-based pyramid of rewards.

Methodology

The methodology deployed by this research projects involves a number of stages. The first stage was rigorous review of the existing literature concerning the conflict diamonds problem and international legal responses to it, as well as applicable literature on regulatory theory. The second stage of the methodology was to draw on the existing regulatory theory literature in order to develop a new theoretical model that might explain how the conflict diamonds governance system might, in a descriptive sense, be analysed and understood, as well as offering insights, in a normative sense, as to how the system might be improved upon.

As part of the first stage of the methodology — collecting information about the conflict diamonds problem and international legal responses to it — a limited exercise in collecting interview data was undertaken. In order to obtain information on the operation of the Kimberley Process, a number of standardised questions were presented to government, NGO, and industry participants in the Kimberley Process. The government participant, the Australian Federal Government, provided a written response while the NGO participant, Global Witness, and the industry participant, Rio Tinto, responded to the questions by means of a verbal interview process. The interviews were recorded and a written transcript produced. In relation to all three participants, the ethical requirements set out by The Australian National University concerning such research were complied with and approved by the ANU Research Ethics Committee.

Road Map

This introduction has set out the two research questions to be explored by the book, namely:

1. To what extent has the conflict diamonds governance system achieved its objectives?
2. Does an application of the networked pyramid regulatory model to the system provide descriptive or normative insights into its effectiveness?

This chapter advances the argument that the conflict diamonds governance system has made modest gains, but has failed in its efforts to address situations of serious non-compliance by member governments. In accordance with insights from the networked pyramid model, the book argues that procedures for expulsion in such cases must be clearly defined and implemented, and that pathways of escalation to the UNSC and the international criminal tribunals should be strengthened through the creation of a new international crime of trafficking in conflict diamonds. It also sets out a road map by summarising in brief the contribution of each chapter towards responding to the research questions, and advancing the main argument of the work. It also provides a more detailed introduction into the substantive chapters of the book.

The second chapter briefly outlines the history and features of the legitimate diamond trade, before turning to the problem of conflict diamonds. It discusses the definition of conflict diamonds, also known as blood diamonds, before detailing the connection of this trade to human rights violations and armed conflict in Angola, Sierra Leone, the DRC, and Côte d'Ivoire.

The third chapter, which discusses the Kimberley Process, is the first of three chapters dealing with the conflict diamonds governance system. It discusses the operation of the Kimberley Process at the international level, including its procedures for accepting new members, monitoring through peer review, and dealing with situations of serious non-compliance. It also considers the particular roles played by NGOs and industry. The fourth chapter is concerned with the domestic implementation of Kimberley responsibilities by national governments.

The UNSC and international criminal tribunals, the other components of the conflict diamonds governance system, are discussed in the fifth chapter. The monitoring role of the UNSC expert committees is discussed, as is its ability to take enforcement action through the imposition of diamond trading sanctions. The track record of the Sierra Leone Special Court and the ICC in prosecuting conflict diamonds cases is then discussed. The book notes that conflict diamonds were discussed in three ways in the emerging jurisprudence: as context for the commission of crimes, as being connected to crimes in the process of mining, and as providing a mechanism of indirect liability between high leadership and direct perpetrators on the ground.

The networked pyramid regulatory model is the subject of the sixth chapter, which discusses the utility of using a regulatory approach, before discussing the features of network models and pyramid models. The networked pyramid hybrid model is then discussed, combining as it does the dialogic and socialisation elements of network models with the ability to ratchet up to more coercive interventions in appropriate circumstances.

Before applying the networked pyramid model, the seventh chapter suggests two major modifications so as to optimise its utility in relation to the conflict diamonds governance system. The first modification, depicting regulation at the national level as a pyramid within the greater international pyramid, attempts to capture the complexity of a regulatory system that operates simultaneously at national and international levels, and in which national governments are both regulators as well as the subjects of regulation. A further modification, showing incentives and sanctions as part of a single model, allows interactions between the two regulatory ratchets to be more clearly observed.

The DNPM is applied to the conflict diamonds governance system in Chapter 8. In this chapter, the two central research questions are discussed in depth and responded to. It is argued that the conflict diamonds governance system has made progress towards its goals, noting that the conflict diamonds trade has reduced to less than 1 per cent of the international diamond trade, and that peace has emerged in Angola and Sierra Leone, both countries that were previously affected by the problem. The contribution of the UNSC and the international criminal tribunals is also noted. The chapter notes, however, the failure of the Kimberley Process to respond to serious non-compliance by three of its government members — Zimbabwe, Angola, and Venezuela — and pyramid theory is recommended as a way forward in these cases. If the KP can extricate itself from its current crises, a further horizon beckons, in which a potential extended mandate might focus on the concept of development diamonds, which are free not only of the taint of conflict and international crime, but a further range of human rights ills. Diamonds mined and polished without the use of child labour, for example, might qualify for voluntary certification, thereby opening a door to fair trade markets in the developed world. Chapter 9, the final chapter, gives a recapitulation of the book, lays out a range of recommendations for possible adoption by parties involved in the conflict diamonds governance system, and suggests areas for further research.

The Conflict Diamonds Problem

In the period following the end of the Cold War, it has become commonplace to observe that the nature of warfare has changed from being predominantly international in character to intranational. The two world wars, paradigms of clashes between nation states, have given way to conflicts between component populations of nations, such as occurred in the early 1990s with the collapse of the former Yugoslavia and the resurgence of ethnically based conflict in Rwanda.[6] Concomitant with these conflicts has been the perpetration of human rights violations of sufficient scale and severity to merit the use of the terminology of international criminal law — namely war crimes, crimes against humanity, and genocide. Another feature of modern conflict has more recently come to the attention of academics and the international community more broadly. This feature is the connection between natural resources and conflict, which has been dubbed the 'resource curse'.[7]

---

6   Dallaire, R, *Shake Hands with the Devil: The Failure of Humanity in Rwanda* (Random House Canada, 2003) 40–41, 516–517; Kofele-Kale, N, 'The Global Community's Role in Promoting the Right to Democratic Governance and Free Choice in the Third World' (2005) 11 *Law and Business Review of the Americas* 205, 215.

7   Le Billon, P, 'The Geopolitical Economy of "Resource Wars"' in P Le Billon (ed.), *The Geopolitics of Resource Wars: Resource Dependence, Governance and Violence* (Frank Cass, 2005) 1–28.

Whether one considers the connection between the oil trade and conflict in Sudan and Iraq, or the association between illegal drugs and warfare in Colombia and Afghanistan, the resource curse has been blamed for the instigation and perpetuation of conflict and gross human rights abuses. Of primary concern has been the fact that belligerent parties, often insurgent groups, have had their armaments funded through proceeds from these commodities.[8] In the context of the African continent, the resource curse has manifested itself through the trade in rough diamonds. Diamonds used to fund the perpetration of warfare and human rights violations by insurgents are now known as conflict diamonds or blood diamonds.[9]

African conflict diamonds were brought to the attention of the international community as a result of the war in Angola, which commenced with independence from Portugal in 1975 and continued until 2002.[10] During the course of the conflict, the rebel group National Union for the Total Independence of Angola (UNITA) took control over all of the major diamond-producing areas of the country. The ceasefire of 1991–92 provided UNITA with the opportunity to sell much of its large harvest of rough diamonds on international markets, using the proceeds to purchase armaments in anticipation of a resumption of the conflict.[11] The non-governmental organisation (NGO) Global Witness at this time was monitoring the situation and, in particular, the annual reports of the South Africa–based diamond mining giant De Beers Corporation. It was well known that De Beers had a standing policy of buying out as much of the global diamond production as it could manage, and the period 1991–92 was no exception. The logical

---

8   Ross, M, 'How Do Natural Resources Influence Civil War?: Evidence from Thirteen Cases' (2004) 58 Winter *International Organization* 35, 35–67; Banat, A B, 'Solving the Problem of Conflict Diamonds in Sierra Leone: Proposed Theories and International Legal Requirements for Certification of Origin' (2002) 19 *Arizona Journal of International and Comparative Law* 939, 939–973.

9   Dunn, K C, 'Identity, Space and the Political Economy of Conflict in Central Africa' in P Le Billon (ed.), *The Geopolitics of Resource Wars: Resource Dependence, Governance and Violence* (Frank Cass, 2005) 151–153, 242–279; Tailby, R, 'The Illicit Market in Diamonds' (2002) 218 *Australian Institute of Criminology: Trends and Issues in Crime and Criminal Justice* 1, 1–6; Saunders, L, 'Note: Rich and Rare are the Gems they War: Holding De Beers Accountable for Trading Conflict Diamonds' (2001) 24 *Fordham International Law Journal* 1402.

10   Price, T M, 'Article: The Kimberley Process: Conflict Diamonds, WTO Obligations, and the Universality Debate' (2003) 12 *Minnesota Journal of Global Trade* 1, 8-11; Kaplan, M, 'Junior Fellows' Note: Carats and Sticks: Pursuing War and Peace through the Diamond Trade' (2003) 35 *New York University Journal of International Law and Politics* 559, 573–578.

11   Price, T M, 'Article: The Kimberley Process: Conflict Diamonds, WTO Obligations, and the Universality Debate' (2003) 12 *Minnesota Journal of Global Trade* 1, 8–11.

conclusion, which was not denied by De Beers, was that diamond sales during the period were going directly or indirectly into the coffers of UNITA, providing funds for the purchase of armaments. By the close of the war in 2002, the conflict had resulted in the loss of up to one million lives.[12]

In 1992, when, for much of the time, the majority of the diamond areas were controlled by UNITA, De Beers stated:

> That we should have been able to buy some two thirds of the increased supply from Angola is testimony not only to our financial strength but to the infrastructure and experienced personnel we have in place.[13]

Significantly, De Beers confirmed the nature of its Angolan business practices in written testimony to a hearing on conflict diamonds held before the United States Congress in May 2000:

> De Beers believes that to regard as 'conflict diamonds' all diamonds emanating from areas of Angola which were from time to [time] under UNITA control during this period [the Angolan civil war] muddles history to make a dubious point. De Beers makes no secret of the fact that during this period it purchased Angolan diamonds on the outside market, although it never at any stage bought diamonds from UNITA itself these purchases were made in good faith and under normal and customary market terms.[14]

Diamonds became associated not only with conflict but also the perpetration of egregious human rights violations in the context of the Sierra Leonean civil war of 1991–2002.[15] The Revolutionary

12   Global Witness, 'A Rough Trade: The Role of Companies and Governments in the Angolan Conflict' (Report, Global Witness Ltd, 1998) 6–8; Price, T M, 'Article: The Kimberley Process: Conflict Diamonds, WTO Obligations, and the Universality Debate' (2003) 12 *Minnesota Journal of Global Trade* 1, 8.
13   De Beers Diamond Jewellers Ltd, 'Annual Report' (1992) cited in Global Witness, 'A Rough Trade: The Role of Companies and Governments in the Angolan Conflict' (Report, Global Witness Ltd, 1998) 8.
14   De Beers Diamond Jewellers Ltd, 'Written Testimony before the United States Congress, House Committee on International Relations, Subcommittee on Africa', *Hearing into the Issue of 'Conflict Diamonds'* (25 May 2000). Available at: www.diamonds.net/fairtrade/Article.aspx?ArticleID=4046. The oral testimony of witnesses appearing before the committee is available at: www.fas.org/asmp/resources/govern/hrgdiamonds.htm. De Beers was not represented in person due to an ongoing legal dispute with the US Government over alleged monopolistic corporate practices. Bieri, F, *From Blood Diamonds to the Kimberley Process: How NGOs Cleaned Up the Diamond Industry* (Ashgate Publishing Ltd, 2010) 46.
15   Price, T M, 'Article: The Kimberley Process: Conflict Diamonds, WTO Obligations, and the Universality Debate' (2003) 12 *Minnesota Journal of Global Trade* 1, 13–16.

United Front (RUF), based away from the coast in the Kono district, was notorious for the practice of amputating the hands and feet of civilians as a technique of intimidating the local population. The RUF was allegedly supported by the government of neighbouring Liberia, and had its activities funded through the exploitation of Sierra Leone's alluvial diamond fields.[16] Large-scale media campaigns organised by groups such as Global Witness highlighted the connection between diamonds sold to consumers in New York and London and the arming of militia in Sierra Leone, leading to consumer boycotts and ultimately to legal action at national and international levels.[17] Particularly noteworthy have been the passage of a series of UNSC resolutions imposing diamond trading prohibitions on Angola, Sierra Leone, and Liberia. A related development was the creation of the Kimberley Process, an effort by government, business, and NGOs to provide a system of import/export licences so as to distinguish the legitimate trade in diamonds from the illegal trade.[18]

The conflict diamonds problem is not only of historical interest, but presents an ongoing and pressing contemporary challenge. As well as the Zimbabwe Marange diamonds issue, ongoing fighting in the DRC, and the recent civil war in Côte d'Ivoire have both been exacerbated by the trade in conflict diamonds. These challenges have proven to be something of a litmus test as to the effectiveness of the new legal mechanisms for the control of conflict diamonds. Côte d'Ivoire has been the site of a conflict bearing a striking resemblance to the Sierra Leone/Angolan precedents, in that insurgent groups largely captured the northern diamond-rich sector of the country. Diamonds in the DRC have proven to be one of a number of resources that have extended conflict by rebels and national governments in that country, particularly in its north-eastern provinces. Unfortunately, human rights abuses have characterised both conflicts to date.

---

16   Ibid 13; Kaplan, M, 'Junior Fellows' Note: Carats and Sticks: Pursuing War and Peace through the Diamond Trade' (2003) 35 *New York University Journal of International Law and Politics* 559, 568–570.

17   Price, T M, 'Article: The Kimberley Process: Conflict Diamonds, WTO Obligations, and the Universality Debate' (2003) 12 *Minnesota Journal of Global Trade* 1, 30–36.

18   SC Res 1173, UN SCOR, 3891st mtg, UN Doc S/RES/1173 (12 June 1998) (Angola); SC Res 1306, UN SCOR, 4168th mtg, UN Doc S/RES/1306 (5 July 2000) (Sierra Leone); SC Res 1343, UN SCOR, 4287th mtg, UN Doc S/RES/1343 (7 March, 2001) (Liberia); Kimberley Process, *The Kimberley Process Certification Scheme* (Core Document, 2002).

Nevertheless, it has been possible to see that the international system in both cases has assisted in denying access to international markets for diamonds originating from these rebel-held regions.[19]

The Conflict Diamonds Governance System

An understanding of the nature of governance responses to the conflict diamonds problem also requires an understanding of the nature of the international diamond trade itself. At the exploration stage, diamond deposits themselves are normally described as being either alluvial or kimberlite in nature. Alluvial diamond deposits are those that, by virtue of an existing or historical river system, have become scattered through the topsoil over a given area and are accessible without the need for expensive or sophisticated diamond mining equipment. By contrast, kimberlite deposits are buried deep below the surface of the earth, are concentrated deposits, and require the use of expensive, sophisticated mining equipment, making it a capital-intensive industry that is less intrinsically vulnerable to the efforts of technologically unsophisticated insurgents. There are four main types of countries that are component parts of the international diamond trade: producer countries involved in mining rough diamonds; rough diamond wholesale trading centres such as the UK and Belgium; polishing/cutting countries that prepare rough diamonds for sale; and market countries where diamond products, such as jewellery, are sold to consumers. It is clear that the diamond industry is highly internationalised and reliant on numerous international trade connections to be effective.[20]

The principal response to the problem of conflict diamonds has been the Kimberley Process. This system emerged through tripartite cooperation between business, government, and NGOs with a view to distinguishing the legitimate diamonds trade from the conflict

19   Final Report Panel of Experts on the Illegal Exploitation of Natural Resources and Other Forms of Wealth of the Democratic Republic of the Congo, UN Doc S/2002/1146 (16 October 2002); Report of the Group of Experts Submitted Pursuant to Paragraph 7 of Security Council Resolution 1584 (2005) Concerning Côte d'Ivoire, UN Doc S/2005/699 (7 November 2005).
20   Global Witness, 'Conflict Diamonds: Possibilities for the Identification, Certification and Control of Diamonds' (Report, June 2000), Section 1 'The Structure of the Diamond Industry'; Price, T M, 'Article: The Kimberley Process: Conflict Diamonds, WTO Obligations, and the Universality Debate' (2003) 12 *Minnesota Journal of Global Trade* 1, 30–36.

diamonds trade. Naturally, when divorced from conflict situations, the diamond industry has the potential to be a powerful driver of economic and social development on the African continent. The Kimberley Process focuses on a system of export certificates, through which participant governments certify the legitimacy of the diamonds at the point of export. Importing governments are mandated to seize unlawful imports and take other action, including domestic prosecution, against non-compliant traders. Certification also allows for statistics to be kept regarding the quantity of rough diamonds traded between countries, allowing estimates of the annual diamond trade.[21]

The Kimberley Process also establishes a system for ensuring that governments take their obligations under the process seriously. In particular, the Kimberley Process includes a Participation Committee and a Monitoring Committee. The Monitoring Committee is mandated to consider annual reports by member nations, as well as organising review visits to nations to assess their compliance with the international system. The Participation Committee is charged with considering applications by countries wishing to join the Kimberley Process, as well as taking punitive action against states in the event of serious non-compliance, where this is evident pursuant to the investigations made by the Monitoring Committee.[22]

Beyond the sphere of the Kimberley Process, and predating its formation, the UNSC has played a pivotal role in combating the trade in conflict diamonds. In response to the role of conflict diamonds in the Angolan conflict, the UNSC passed Resolution 1173 of 1998, which was the first international trade ban on the diamond trade, in response to the link of that trade with conflict. The UNSC further intervened to impose trade bans on both Sierra Leone and Liberia, the latter when it came to light that it was being used as a conduit country for smuggling diamonds of Sierra Leonean provenance. The UNSC has also passed resolutions concerning conflict diamonds in response to the situation in the DRC, and, significantly, demonstrated its capacity

---

21 Kimberley Process, *The Kimberley Process Certification Scheme* (Core Document, 2002).
22 Ibid.

to respond in a collaborative fashion by imposing sanctions in 2005 after the Kimberley Process had imposed a trade ban on Côte d'Ivoire diamonds.[23]

A final response to the problem of conflict diamonds has come from the emergent institutions of international criminal justice. The earliest precedents of financial contribution to the perpetration of international crimes goes back to the Nazi war crimes cases of *Flick, Farben and Krupp* at Nuremberg in the aftermath of World War Two.[24] The jurisprudence of 'joint criminal enterprise' developed by the Yugoslavia and Rwanda Tribunals since the mid-1990s now provides a coherent legal framework for bringing such individuals to account.[25] While the ICC, established under the 1998 *Rome Statute,* has alluded to the possibility of persons complicit in the conflict diamonds trade being brought to account, the most important breakthrough has been made by the Special Court for Sierra Leone, which has put on trial the former president of Liberia, Charles Taylor, for his role in the conflict diamonds trade as it related to the Revolutionary United Front (RUF).[26]

The Networked Pyramid Regulatory Model

Concurrent with the recognition of the challenge posed by the conflict diamonds trade, there have been trends within academia to develop models for evaluating the effectiveness of legal systems for

---

23   SC Res 1173, UN SCOR, 3891st mtg, S/RES/1173 (12 June 1998) (Angola); SC Res 1306, UN SCOR, 4168th mtg, S/RES/1306 (5 July 2000) (Sierra Leone); SC Res 1343, UN SCOR, 4287th mtg, S/RES/1343 (7 March, 2001) (Liberia); SC Res 1457, UN SCOR, 4691st mtg, UN Doc S/RES/1457 (24 January 2003) (Democratic Republic of Congo); SC Res 1643, UN SCOR, 5327th mtg, UN Doc S/RES/1643 (15 December 2005) (Côte d'Ivoire).
24   *Trials of Nazi War Criminals before the Neurnberg Tribunals under Control Council Law No 10: Neurnberg October 1946–April 1949, Volumes V, VI, VII, VIII, IX* (United States Government Printing Office, 1952). Reproduced at: www.phdn.org/archives/www.mazal.org/NMT-HOME.htm.
25   *The Prosecutor v Dusko Tadic (Appeal Judgement)* (International Criminal Tribunal for the Former Yugoslavia, Appeals Chamber, Case No IT-94-1-A, 15 July 1999) 227–228; *The Prosecutor v Brdjanin (Appeal Judgement)* (International Criminal Tribunal for the Former Yugoslavia, Appeals Chamber, Case No IT-99-36-A, 3 April 2007) 110–141.
26   International Criminal Court, Office of the Prosecutor, *The Prosecutor on the Co-operation with Congo and other States Regarding the Situation in Ituri, DRC* (26 September 2003); 'Prosecution's Second Amended Indictment', *The Prosecutor v Charles Taylor* (Special Court for Sierra Leone, Case No SCSL-03-01-PT, 29 May 2007); Transcript of Proceedings, 'Evidence of Expert Witness Ian Smillie', *The Prosecutor v Charles Taylor (Trial)*, (Special Court for Sierra Leone, Trial Chamber II, Case No SCSL-03-01-PT, 7 January 2008).

responding to such challenges. Such approaches often go beyond a purely legal analysis into the area of regulatory theory. Regulatory theory is distinguished from legal analysis by assessing whether legal systems actually achieve desired social outcomes, and involves the development of models that, when applied, can increase the ability of systems to achieve their desired outcomes.[27]

Two models from the field of regulatory theory, the network model and the pyramid model, stand out for their potential application in relation to the conflict diamonds governance system. Network models, such as the regulatory web or the horizontal government network, emphasise regulatory techniques based on persuasion, dialogue, and socialisation. These models provide for dynamic interaction with non-governmental participants such as business organisations and civil society organisations, and encompass organisations such as the International Labour Organization, which has formalised but largely private processes of naming and shaming, which create a socialisation pressure tending towards normative compliance.[28]

Another model, the regulatory pyramid, was first developed by Ayres and Braithwaite to explain the ways in which regulatory standards for a particular industry may best be enforced when considered within a particular national jurisdiction. The model is useful in that it does not take an either/or approach to the two conflicting schools of regulatory thought: industry self-regulation, and command-and-control regulation. The industry self-regulation school seeks to use soft compliance options involving the engagement of industry to promote best practice outcomes. Rather than imposing a regime from an external source on a particular industry, such approaches seek to empower the industry itself to recognise its own interest in complying with particular regulatory goals. By contrast, so-called command-and-control systems are imposed from government entities beyond the industry itself, using instruments such as criminal prosecution to forcefully bring industry players to account against government legal standards. The regulatory pyramid model seeks to marry these two seemingly disparate systems by suggesting that

27   Black, J, 'Critical Reflections on Regulation', *Australian Journal of Legal Philosophy*, vol. 27, 2002, 20.
28   Slaughter, A-M, *A New World Order* (Princeton University Press, 2004) 11, 14, 19, 21, 24, 56, 147, 156, 261, 262.

industry self-regulation approaches should be implemented as far as practicable in the absence of command-and-control. As long as this approach based on soft regulation is successful, there is no imperative to change the regulatory dynamic. However, the sensible external regulator will be engaged in careful monitoring of the system for signs of disrepair, at which point sanctions can be ratcheted up and command-and-control can take over as the central regulatory approach. Assuming heavier sanctions such as civil sanctions or fines are successful, and new signs of responsiveness are observed within the industry, sanctions might be ratcheted back down to the status of self-regulation. In the event that compliance is still not forthcoming, more severe strategies might be resorted to, such as criminal sanctions or the dissolution of the corporation. The keystone of the pyramid is that its strategy is to be contingently punitive or forgiving.[29]

More recent thinking in regulatory theory has sought to bring together the salient features of both network and pyramid models to create so-called networked pyramid hybrid models. The networked pyramid hybrid combines network features, such as expanding beyond government players and the focus on techniques of dialogue, persuasion, and socialisation, with pyramid features, which allow for escalation to more coercive interventions where dialogue has proven ineffective. Such modelling has been applied to international regulatory systems, such as the regulatory export of intellectual property standards from the US to other national governments, the field of traditional knowledge as well as the development of the threat of collective debt default by developing countries as a regulatory weapon of the weak.

Adaptation and Application of the Networked Pyramid

Before applying the networked pyramid model to the conflict diamonds governance system, the book suggests that it might be modified in a number of ways for optimal effect. Firstly, a complex system such as this would benefit from a model that could describe regulatory action

---

29   Ayres, I and J Braithwaite, *Responsive Regulation: Transcending the Deregulation Debate* (Oxford University Press, 1992) 20–51.

at international and national levels, where national governments are regulators as well as being regulated. As such, a diagram involving a smaller series of national pyramids at the base of a larger international pyramid is suggested. Furthermore, it is suggested that a regulatory model could benefit from the combination of incentives and sanctions in a single diagram, particularly with reference to the interaction between both sets of interventions. In summary, the revised model is a dual networked pyramid, with the duality being found both in the fact that there is a pyramid within the pyramid, and the combination of sanctions and incentives.

Applying the networked pyramid model to the conflict diamonds governance system provides insights into the analytical power of the model and the areas of potential improvement that might be made to the conflict diamonds system. It is useful, in the first instance, to consider whether the conflict diamonds governance system corresponds descriptively with either or both of the two component models — that is, the degree to which it already embodies the features of a regulatory pyramid or a network. In the event that the legal system is not a perfect fit on the descriptive level, the question to be considered is what changes would need to be made to the system to make a better correspondence with the theoretical model. In the event that these changes would appear to make the conflict diamonds system a more effective system, the models might be considered to be normatively powerful.

The models may be of particular utility in refining the relationship between regulatory actors in the conflict diamonds system, where there is little or no formal institutional linkage, for example between the KP, the UNSC, and the ICC. However, one of the particular challenges in considering the conflict diamonds legal system is that it operates both at the national or domestic level, where the apex of the system is the national government, and at the international level, where the principal regulators are other national governments and international agencies. As such, it may be that a particular model being considered may be a good fit at the domestic level but not at the international level.

While being based primarily on an analysis of existing literature in the fields of conflict diamonds and regulatory theory, the book has also involved a limited exercise in the gathering of primary data. A representative from the business community, a government representative and an NGO representative were all interviewed in relation to their own views on the effectiveness of the conflict diamonds legal system at tackling the illicit diamonds trade. The interviews were carried out according to standard university ethical procedures, and the qualitative results from the interviews have been incorporated into the body of the book.

Concluding Remarks

This chapter has sought to introduce the book and provide an overview. After some preliminary remarks, the two research questions were set out:

1. To what extent has the conflict diamonds governance system achieved its objectives?
2. Does an application of the networked pyramid regulatory model to the system provide descriptive or normative insights into its effectiveness?

The chapter also set out the central argument of the book, that the conflict diamonds governance system has been largely successful, but has failed in its efforts to address situations of serious non-compliance by member governments. In accordance with insights from the networked pyramid model, the book argues that procedures for expulsion in such cases must be clearly defined and implemented, and that pathways of escalation to the UNSC and the international criminal tribunals should be strengthened through the creation of a new international crime of trafficking in conflict diamonds. The chapter also delineated a road map for the book, summarising in brief the contribution of each chapter towards responding to the research questions, and advancing the main argument of the book. It also provided a more detailed introduction into the substantive chapters of the work.

2

Are Conflict Diamonds Forever?: Background to the Problem

There were more than thirty boys there, two of whom, Sheku and Josiah, were seven and eleven years old … 'It seems that all of you have two things in common', the soldier said after he had finished testing all of us. 'You are afraid of looking a man in the eye and afraid of holding a gun.'

Ishmacl Bcah, former child soldier with the RUF[1]

Chapter Overview

This chapter discusses the nature and parameters of the international legitimate diamond trade, before distinguishing it from the conflict diamonds trade. The definition of conflict diamonds, also known as blood diamonds, is discussed. The chapter discusses the role of conflict diamonds in exacerbating armed conflict and human rights violations in Angola, Sierra Leone, the Democratic Republic of Congo (DRC), and Côte d'Ivoire, before giving brief concluding remarks. A strong understanding of the nature of the conflict diamonds problem is essential to any meaningful attempt to respond to the two research questions, which seek to evaluate the effectiveness of the global response to the problem, and how this response might be improved, with reference to the networked pyramid regulatory model.

---

1   Beah, I, *A Long Way Gone: Memoirs of a Boy Soldier* (Sarah Chrichton Books, 2007).

The Mainstream Diamond Trade

The mainstream trade in rough or unprocessed diamonds is a multinational, multibillion dollar industry that, until very recently, has resisted modern trends towards transparency in its dealings. Diamond production in the legal industry during 2006 was valued at US$11.5 billion, representing 176.7 million carats, where each carat is 0.2 grams in weight. Global diamond imports were valued at US$37.7 billion, representing 500.5 million carats, and exports were valued at US$38.4 billion, representing 514.2 million carats.[2]

The diamond industry has been dominated by the De Beers corporation for more than 100 years. De Beers produces about half the world's rough diamonds, calculated by value, and regulates world prices for unprocessed stones by purchasing and stockpiling up to 80 per cent of the world's rough diamond output.[3] Diamonds are sorted in London into approximately 5,000 categories by size and quality. Most diamonds are then distributed to dealers in Antwerp, where the majority of rough diamond trading occurs. Other major centres are London, Lucerne, New York, Tel Aviv, Johannesburg, Bombay, and Dubai. The cutting and polishing of diamonds occurs in approximately 30 countries, including India, South Africa, Botswana, Russia, China, Sri Lanka, Thailand, Vietnam, and Mauritius. A diamond would have been traded several times before arriving at one of the major jewellery-making centres located in Israel, Belgium, India, and New York.[4]

Connected to the virtual monopoly exercised by De Beers, diamond transactions have neither been subject to the rigour of tough competition, nor strictly regulated by governments. Reliable statistics regarding the quantity and value of the rough diamond trade in recent decades are hard to come by. In 1998, for example, the Government of Sierra Leone recorded that 8,500 carats of diamonds were exported to Belgium, whereas records in Belgium indicated that 770,000 carats

---

2   Kimberley Process Secretariat, *Annual Global Summary: 2009 Production, Imports, Exports and KPC Counts* (Annual Report Summary, 7 August 2010).

3   Global Witness, 'A Rough Trade: The Role of Companies and Governments in the Angolan Conflict' (Report, Global Witness Ltd, 1998) 5–6.

4   Global Witness, 'Conflict Diamonds: Possibilities for the Identification, Certification and Control of Diamonds' (Report, June 2000), Section 1 'The Structure of the Diamond Industry'; Price, T M, 'Article: The Kimberley Process: Conflict Diamonds, WTO Obligations, and the Universality Debate' (2003) 12 *Minnesota Journal of Global Trade* 1, 30–36; Transcript of Proceedings, 'Evidence of Expert Witness Ian Smillie', *The Prosecutor v Charles Taylor (Trial)*, (Special Court for Sierra Leone, Trial Chamber II, Case No SCSL-03-01-PT, 7 January 2008) 522.

were imported that year from Sierra Leone. The discrepancy may be attributed to a combination of diamonds being exported without the knowledge of Sierra Leone, and also a tendency in Belgium not to rigorously investigate information provided to it regarding the country of export. To take another example, a United Nations Security Council (UNSC) investigation found that the addresses of Liberian companies appearing on Liberian export invoices in Antwerp, when investigated in Liberia, did not house actual company offices. In the case that postal mail was directed to the company, courier companies had been instructed to redirect the mail to the Liberian International Shipping and Corporate Registry.[5]

Identifying a country of origin for particular diamonds was made harder by the fact that the Antwerp record-keeping system, as well as other systems, identified only the country of last export, rather than the country of origin of the diamonds, thereby obscuring attempts to discriminate between the legitimate and black-market trades.[6] The initial reluctance to confront the conflict diamonds problem by industry players can largely be attributed to this lack of transparency, and consequent reluctance to share information freely. It might be noted that the illegal trade in diamonds includes not only the conflict diamonds trade, but also other forms of smuggling aimed at tax avoidance or money laundering.[7]

The success of the diamond industry has largely ridden on its advertising approaches. Diamonds have become the most legitimate and acceptable symbol of marital engagement. In the 1930s, De Beers promulgated this image to recover dwindling sales in the Great Depression. Eighty per cent of engagements in the United States were consecrated with diamond rings by 1950, after the Diamonds Are Forever campaign. The United States and Europe are the largest consumer markets for diamond jewellery, representing about 65 per cent of the global market, while the demand from the Chinese market is rapidly expanding.[8]

---

5   Transcript of Proceedings, 'Evidence of Expert Witness Ian Smillie', *The Prosecutor v Charles Taylor (Trial)*, (Special Court for Sierra Leone, Trial Chamber II, Case No SCSL-03-01-PT, 7 January 2008) 543–544.

6   Smillie, I, L Gberie and R Hazleton, 'The Heart of the Matter: Sierra Leone, Diamonds and Human Security' (Report, Partnership Africa Canada, 2000) 6.

7   Saunders, L, 'Note: Rich and Rare are the Gems They War: Holding De Beers Accountable for Trading Conflict Diamonds' (2001) 24 *Fordham International Law Journal* 1402, 1414.

8   Koyame, M, 'United Nations Resolutions and the Struggle to Curb the Illicit Trade in Conflict Diamonds in Sub-Saharan Africa' (2005) 1 *African Journal of Legal Studies* 80, 94–95.

A product built on positive publicity can, however, fall by negative publicity focused on the association of diamonds with conflict and human rights violations. In 1998, Global Witness thrust the conflict diamonds problem into the public arena by highlighting its connection with the Angolan conflict. Protests in New York outside Tiffany & Co. jewellers led to publicity in the *New York Times*. Aware of the effect of negative publicity in debilitating the fur coat trade, media coverage of the conflict diamonds issue has been an important force in galvanising the diamond industry to take the issue seriously. It has also been argued that the existing diamond industry, which is still largely dominated by De Beers, has a strong interest in enforcing the Kimberley Process (KP), as it is a further means of shoring up a virtual monopoly over the international diamond trade.[9]

In 2002, business, NGOs, and governments combined to create the Kimberley Process Certification Scheme (KP), aimed at tackling the problem of conflict diamonds. The thrust of the Kimberley Process, discussed further in Chapter 4, is to create a paper trail between diamond miners at the beginning of the diamond pipeline, and end consumers of diamond products, so as to distinguish between the legal diamond trade and the illegal trade, thereby preventing the sale of conflict diamonds. Such a system is necessary to identify the country of origin of diamonds, as there is currently no available technology that can accurately identify the country of origin of a diamond simply through analysis of the stone. At most, generalisations based on the value of the stone might be made, noting, for example, that Sierra Leone typically produces stones valued at about US$200 per carat, as compared to Canadian diamonds valued at about US$100 per carat, or Congolese diamonds valued at about US$25 per carat. Diamond value is assessed based on the qualities of clarity, colour, carats, and cut.[10]

---

9   Price, T M, 'Article: The Kimberley Process: Conflict Diamonds, WTO Obligations, and the Universality Debate' (2003) 12 *Minnesota Journal of Global Trade* 1, 30–36; Black, B, 'Panel: Combating International Corruption through Law and Institutions' (2007) 5 *Santa Clara Journal of International Law* 445, 461; Holmes, J, 'The Kimberley Process: Evidence of Change in International Law' (2007) 3 *Brigham Young University International Law and Management Review* 213, 218–219.

10   Transcript of Proceedings, 'Evidence of Expert Witness Ian Smillie', *The Prosecutor v Charles Taylor (Trial)*, (Special Court for Sierra Leone, Trial Chamber II, Case No SCSL-03-01-PT, 7 January 2008) 518, 534.

Naturally, the legitimate diamond trade holds great potential for the economic and social development of African producer nations, as is implied in the term development diamonds.[11] If diamond revenues were to benefit the country's population, and where an appropriate amount is paid as taxation revenue, the industry could become a constructive force. Unlike other trades, such as the trade in heroin, cocaine, and other debilitating drugs, it is hard to argue that there is anything intrinsically unethical about trading in diamonds. It is their connection with human rights abuses and conflict that makes black market trading a pariah. An interesting parallel can be made between the trade in certified diamonds and the trade in antique ivory products under the *Convention on the International Trade in Endangered Species*. In certain circumstances, the ivory trade is undesirable, namely where the trade in fresh ivory, or products made from it, is allowed. Naturally, this trade encourages the killing of elephants, an endangered species under the convention. The sale of antique items made using existing stocks of ivory, however, is arguably a distinct market, which does not require the continued killing of elephants.[12]

Conflict Diamonds

The definition of conflict diamonds used in the context of international law is that found in the Kimberley Process core document, which is based on Resolution 55/56 of the United Nations General Assembly (UNGA), and relevant UNSC resolutions:

> Conflict diamonds [are] rough diamonds used by rebel movements or their allies to finance conflict aimed at undermining legitimate governments.

While it is important to find a workable definition for conflict diamonds, the definition that was arrived at in these early resolutions, and that found its way into the Kimberley Process core document, is open to some criticism. The definition aims to capture part of the concept of conflict diamonds, namely the role that the black market in

---

11 'Development diamonds' are discussed at length in Chapter 3.
12 For a useful discussion of the Convention on the International Trade in Endangered Species, see United Kingdom Wildlife Licensing and Registration Service, 'Guidance for Antique Dealers on the Control of Trade in Endangered Species' (2005). Available at: www.culturecommunication. gouv.fr/content/download/97744/875972/version/1/file/2005_Guidance+for+Antique+dealers +on+the+control+of+trade+in+endangered+species.pdf.

diamonds plays in fuelling warfare. However, in focusing solely on the element of warfare, the definition fails to identify the important link to the separate concept that the black market in these circumstances also fuels serious human rights violations. It is a premise of international humanitarian law, the human rights conventions that apply during times of conflict, that warfare is not intrinsically illegal. Warfare only becomes illegal when fundamental principles are violated, such as the principle of distinction, which distinguishes between military personnel, who may be legitimately attacked, and others, including civilians and wounded soldiers, who may not be attacked.[13] Perhaps the key element in harnessing world opinion against the trade in conflict diamonds has been its connection with serious violations of human rights, including the principle of distinction. The wars in Sierra Leone and Angola, for example, have both been characterised by the targeting and killing of civilians.

A further difficulty with the international definition of conflict diamonds is its differentiation between rebel movements and legitimate governments. One of the defining features of the development of international law during the past few decades has been a formal recognition that parties in a non-international conflict (i.e. rebel movements) are bound by the same laws of warfare as parties to the more established category of conflict between national armies.[14] The terminology used in this context also suggests that recourse to warfare by rebel movements is always in contravention of international law. International law, however, recognises that recourse to warfare may be justified in certain circumstances, including a war of self-defence, and wars against a colonising power.

The distinction made between rebel movements and legitimate governments also suggests that governments may legally fund their wars through the diamond trade, and, more problematically, the implication that government forces by their very nature do not commit human rights abuses. Although it may be difficult to outlaw the manner in which the diamond trade finances wars fought by

---

13   For a useful discussion on the principle of distinction, see International Committee of the Red Cross, *Rule 1. The Principle of Distinction between Civilians and Combatants.* Available at: www.icrc.org/customary-ihl/eng/docs/v1_ru_rule1.

14   *Protocol Additional to the Geneva Conventions of 12 August 1949, and relating to the Protection of Victims of Non-International Armed Conflicts (Protocol II)*, opened for signature 8 June 1977, 1125 UNTS 609 (entered into force 7 December 1978).

national governments, it is particularly problematic to suggest that government armies do not commit human rights violations. In all of the conflicts discussed in this section — Angola, Sierra Leone, the DRC, Côte d'Ivoire, and Zimbabwe — reports of serious human rights violations have been made not only in relation to rebel movements but also government armies. Furthermore, during the conflict in the DRC, according to a key report by the UNSC Expert Committee, the armies of Uganda, Rwanda, and Zimbabwe were all mining Congolese natural resources, including diamonds, to further their war efforts. The issue has also been a thorn of contention in relation to Zimbabwean rough diamonds originating from the Marange diamond fields. It entails alleged human rights abuses committed by Zimbabwean authorities against alluvial miners at Marange. As such, it does not involve either a civil war or an international conflict, and would not, on its face, fall within the conflict diamonds definition as it appears in the Kimberley core document.

The connection between the trade in black market diamonds and human rights violations is perhaps better expressed by the term blood diamonds than conflict diamonds. Although the terms are used interchangeably, the connection with blood arguably connotes the violence directed against civilians better than the more prevalent term conflict diamonds. Perhaps the most infamous example of the connection between this trade and gross human rights violations is the recent civil war in Sierra Leone. The Revolutionary United Front (RUF) militia have been documented as committing crimes of terror to subdue civilian populations, including the amputation of hands and limbs. Weapons and other resources that supported these activities were largely funded through the occupation of diamond mining areas by the RUF, allegedly assisted by the Liberian Government.[15]

Considering the challenges with the definition of conflict diamonds, the non-legal status of the Kimberley Process may again prove to be a benefit. Such an issue is more problematic in legal status documents that are subject to established norms of interpretation. In the absence of a legally binding approach, the Kimberley Process has been able to take a broader interpretation of the definition to encompass rough

---

15   Saunders, L, 'Note: Rich and Rare are the Gems They War: Holding De Beers Accountable for Trading Conflict Diamonds' (2001) 24 *Fordham International Law Journal* 1402, 1402–1428; Tailby, R, 'The Illicit Market in Diamonds' (2002) 218 *Australian Institute of Criminology: Trends and Issues in Crime and Criminal Justice* 1, 1–6.

diamonds originating from Zimbabwe's Marange Field. Influential commentators, such as Ian Smillie, have argued for a 'purposive' interpretation of the Kimberley conflict diamonds definition.[16] In particular, Smillie points to the human rights reference in the perambulatory passages of the Kimberley core document:

> Recognising the devastating impact of conflicts fuelled by the trade in conflict diamonds on the peace, safety and security of people in affected countries and the systematic and gross human rights violations that have been perpetrated in such conflicts.[17]

With reference to this statement, he argues that the Kimberley Process has always been concerned 'about the appalling human rights abuses committed in the course' of conflicts. A purposive interpretation is even available with reference to 'black letter' international law. The *Vienna Convention on the Law of Treaties*, article 31, states:

> A treaty shall be interpreted in good faith in accordance with the ordinary meaning to be given to the terms of the treaty in their context and in the light of its object and purpose. [18]

If customary international law is to be invoked, evidence of 'state practice' must refer to the fact that the UNGA accepted Marange diamonds as conflict diamonds in a number of their resolutions.

Recognising the political importance, if not the legal necessity, of reinforcing the broader definition of conflict diamonds, Smillie recommends an amendment to the Kimberley core document. He suggests that the following wording be added:

> The Kimberley Process promotes respect for human rights as described in the Universal Declaration of Human Rights, and it requires their effective recognition and observance, as part of KPCS minimum standards, in the diamond industries of participating countries, and among the peoples, institutions and territories under their jurisdiction.[19]

---

16   Smillie, I, 'Paddles for Kimberley: An Agenda for Reform' (Report, Partnership Africa Canada, June 2010) 15–16.
17   Kimberley Process, *The Kimberley Process Certification Scheme* (Core Document, 2002) 1.
18   *Vienna Convention on the Law of Treaties*, opened for signature 23 May 1969, 1155 UNTS 331 (entered into force 27 January 1980) art 31.
19   Smillie, I, 'Paddles for Kimberley: An Agenda for Reform' (Report, Partnership Africa Canada, June 2010) 16.

It might be argued, however, that a more explicit amendment to the definition of conflict diamonds would be better in the interests of promoting certainty. Rather than adding a general reference above to the recognition of human rights, it might be advisable to clarify the definition of conflict diamonds, perhaps in terms such as these:

> Conflict diamonds are rough diamonds, the production of which is associated with, or the sale of which finances, the commission of international crimes, including war crimes, genocide and crimes against humanity.

The cause of the Kimberley Process would be politically and legally reinforced if the clarified definition was endorsed by resolutions of the UNGA and the UNSC.

It would appear, however, that clarifications to the KP mandate have already met with political opposition within the KP Plenary. At the 2010 KP Plenary, the KP Civil Society Coalition introduced, for the fourth straight meeting, language seeking to clarify the relationship between the KP and human rights. The language stated that KP participants should respect international human rights law when providing security in their diamond sectors. Civil Society, supported by the World Diamond Council and a majority of governments, argued that the credibility of the KP would be seriously undermined if it was not seen to be actively engaged in preventing and responding to human rights violations by state agents in the diamond sector. Despite this support, consensus was blocked by India, China, Russia and the DRC. Botswana and Namibia reserved judgement, saying they needed more time to study the initiative.[20]

While the Sierra Leone and Angolan wars fuelled by diamonds have now ended, diamonds still fuel conflict in the north-eastern Ituri region of the DRC, as well as Côte d'Ivoire. The world's three largest UN peacekeeping forces are in Sierra Leone, Liberia, and the DRC, consisting of 35,000 troops, with combined budgets of $1.8 billion.[21] There is also a documented link between conflict diamonds and

---

20   Partnership Africa Canada, *Other Facets: News and Views on the International Effort to End Conflict Diamonds*, No. 34 (February 2011) 4.
21   Global Witness and Partnership Africa Canada, 'Rich Man, Poor Man, Development Diamonds and Poverty Diamonds: The Potential for Change in the Artisanal Alluvial Diamond Fields of Africa' (Report, October 2004) 3.

international terrorist groups such as Al Qaeda. Grave human rights violations that, on their face, constitute international crimes, continue to be committed on the artisanal diamond fields of Zimbabwe and Angola by government forces.[22]

Diamonds possess qualities that lend themselves to the exacerbation of conflict. They are easy to mine without complex equipment, particularly where there is an abundance of manual labour, which makes the miners an easy target for militia groups. Alluvial diamonds are the most vulnerable, being diamonds distributed close to the surface of the earth as a result of being moved by existing or historical river systems. Miners of alluvial diamonds are often called artisanal miners, because they are able to do the mining using only a shovel and a sieve. Artisanal miners also take advantage of tailings, such as those found in Sierra Leone, which are deposits of diamondiferous gravel that have been abandoned in the wake of large-scale industrial mining. Alluvial diamonds can be contrasted with kimberlite diamond deposits, which are concentrated deposits embedded deeply beneath the earth's surface, normally accessible only with the use of sophisticated mining equipment.[23]

Gem-quality diamonds have historically held their value well, which makes them a good investment and useful as a form of hard currency to launder money, purchase weapons, or stockpile for later use. They are the world's most concentrated form of wealth, being very small and of high value, which makes them easy to transport or smuggle. They do not show up on a standard metal detector, although they would be detectable by an x-ray machine. The unregulated nature of the diamond industry has, until recently, contributed to the problem, as there have been few trading restrictions and the legal industry has traditionally not been transparent in its dealings. Multiple transactions,

---

22   Global Witness, 'For a Few Dollars More: How Al Qaeda Moved into the Diamond Trade' (Report, April 2003) 8–9.

23   Transcript of Proceedings, 'Evidence of Expert Witness Ian Smillie', *The Prosecutor v Charles Taylor (Trial)*, (Special Court for Sierra Leone, Trial Chamber II, Case No SCSL-03-01-PT, 7 January 2008) 517–520, 524–525, 530–531; Global Witness and Partnership Africa Canada, 'Rich Man, Poor Man, Development Diamonds and Poverty Diamonds: The Potential for Change in the Artisanal Alluvial Diamond Fields of Africa' (Report, October 2004) 3.

international transfers, and the practice of mixing diamonds from different sources obscure the origin of diamonds, thereby facilitating smuggling and illegal behaviour.[24]

Beyond the connection between rough diamonds and international human rights crimes lies a further range of issues that problematise the diamond industry. International crimes are those human rights violations that are considered the most serious under international law, with the technical description of war crimes, crimes against humanity, and genocide, which fall within the subject matter jurisdiction of international criminal tribunals such as the International Criminal Court (ICC). Human rights violations such as murder, rape, the recruitment and use of child soldiers, or forced labour, where carried out on a widespread or systematic basis against civilians, are examples of such crimes. Other human rights violations, while still considered serious, are not classified as international crimes under international law. This second category includes child labour, such as parents including their children in artisanal mining activities, and violations of International Labour Organization conventions such as those relating to health and safety, and minimum levels of remuneration for labour.

Lower level human rights violations do not fall within the current mandate of the Kimberley Process, however, there is a clear connection between these problems and the risk of escalation to the commission of international crimes. Put differently, well regulated, healthy, and safe artisanal mining communities are less likely to be attracted or co-opted to sell their proceeds on the black market to the benefit of rebel militias. As a result, these lower level issues are discussed in the country-by-country section. Chapter 6 suggests a framework for extending the Kimberley Process mandate to encompass these lower level violations. One of the countries discussed, India, does not have an international crime level issue, but has other human rights issues associated with its diamond cutting and polishing centres, particularly the use of child labour. This information is included with the view in mind that the Kimberley Process mandate might at some stage take into account this broader range of human rights issues.

---

24   Transcript of Proceedings, 'Evidence of Expert Witness Ian Smillie', *The Prosecutor v Charles Taylor (Trial)*, (Special Court for Sierra Leone, Trial Chamber II, Case No SCSL-03-01-PT, 7 January 2008) 522–523; Global Witness, 'For a Few Dollars More: How Al Qaeda Moved into the Diamond Trade' (Report, April 2003) 1–3.

Estimating the Size of the Conflict Diamonds Trade

Although the period prior to the establishment of the Kimberley Process was characterised by secrecy and a lack of transparency, particularly in the area of statistics, nevertheless, attempts were made by several organisations to assess the overall size of the rough diamond trade, and the percentage of that trade represented by the trade in conflict diamonds. On the conservative side, De Beers estimated that, in 1999, conflict diamonds represented 3.7 per cent of the world rough diamond trade. The source countries for conflict diamonds in that year, according to De Beers, were Angola, Sierra Leone, and the DRC. It might be noted, however, that De Beers was adhering to the so-called narrow definition of conflict diamonds, namely, that they had to be fuelling rebel militias against legitimate governments. It was presumably on this basis that the entirety of Angola's rough diamond production was not included in the statistic, so as to exclude Angolan government rough diamonds from the conflict diamonds equation. According to the broad reading of the conflict diamonds definition, all Angolan diamonds that year should have been classified as conflict diamonds, given the fact that both parties to the civil war committed international human rights crimes.[25]

In its three-year review, the Kimberley Process estimated that the percentage of conflict diamonds trade was thought to have been in the range of 4 to 15 per cent between the mid-1990s and the beginning of the 2000s.[26]

Turning to the estimates of what the conflict diamonds trade may amount to right now, it is a particularly contested issue in light of the fact that participants within the Kimberley Process have split over the definition of conflict diamonds. As some national governments have preferred the narrow definition, which confines conflict diamonds to a connection with civil war rather than international human rights

---

25   De Beers Diamond Jewellers Ltd, 'Written Testimony before the United States Congress, House Committee on International Relations, Subcommittee on Africa', *Hearing into the Issue of 'Conflict Diamonds'* (25 May 2000). Available at: www.diamonds.net/News/NewsItem.aspx?ArticleID=4046.

26   Kimberley Process, *The Kimberley Process Certification Scheme: Third Year Review,* Kimberley Process (November 2006) 17. Available at: www.state.gov/documents/organization/77156.pdf.

crimes, rough diamonds originating from the Marange artisanal fields of Zimbabwe and the artisanal fields of northern Angola have not been classified as conflict diamonds. By contrast, others consider that the definition of conflict diamonds incorporates a connection between the mining and trade of rough diamonds with the commission of international human rights crimes. Under this broad definition, the association of Zimbabwean and Angolan rough diamonds with international human rights crimes means that these diamonds must be classified as conflict diamonds.

The three-year review of the Kimberley Process estimated that conflict diamonds represented about 0.2 per cent of the world's rough diamond trade in 2004. The estimate was based around conflict diamonds production from Côte d'Ivoire and UNSC embargoed diamonds from Liberia, although the figure did not take into account the ongoing fighting centred on diamond mines in the DRC. While the official Kimberley Process organs continue to refer to this figure, those who take the broader definition of conflict diamonds recognise that this figure was calculated without reference to Angolan or Zimbabwean conflict diamonds and is likely to contain further inaccuracies.[27]

A Country by Country Approach

Angola

The Angola conflict was the first to bring international attention to the problem of conflict diamonds. The conflict had its genesis when Angola was granted independence from its former colonial master, Portugal, in 1975. While the Popular Movement for the Liberation of Angola (MPLA) was recognised as the first independent Angolan government, it was resisted throughout the 1980s and 1990s by the National Union for the Total Independence of Angola (UNITA) and its ally, the National Liberation Front of Angola (FNLA). [28] There were reports of gross human rights violations on both sides, including the

---

27   Ibid.
28   Maggi, M, 'The Currency of Terrorism: An Alternative Way to Combat Terrorism and End the Trade of Conflict Diamonds' (2003) 15 *Pace International Law Review* 513, 522.

indiscriminate shelling of civilians.[29] The war resulted in the loss of up to 1 million lives, with 1.4 million in need of food aid, 500,000 in critical danger of starvation, and the country burdened with 4–5 million land mines, killing or injuring 700 Angolans per year.[30]

In 1992 there was a resurgence of the conflict following UN monitored elections that confirmed the legitimacy of the MPLA Government. UNITA, led by Jonas Savimbi, refused to accept the result and returned to civil war, focusing on control of the diamond-producing areas of Angola.[31] Diamonds were needed to fund the conflict, given that the end of the Cold War had resulted in the loss of financial backing from the United States and South Africa.[32] UNITA either directly exploited diamond mining areas, or used systems of taxation and licensing to extract commission from the labour of others. Proceeds from diamond sales were then used to purchase weapons.[33] Diamonds were also an important component of UNITA's strategy for acquiring friends and maintaining external support. UNITA gained particular support from the Mobutu Government in what was then Zaire.[34] Rough diamond caches, rather than cash or bank deposits, also constituted the primary and the preferred means of stockpiling wealth for UNITA. This provided a mechanism for avoiding the effects of international financial sanctions, such as confiscation of bank accounts, and provided a way of sustaining income expenditure over a long period.[35]

In response to the renewed violence, the UNSC imposed a mandatory embargo on the sale or supply of weapons or petroleum products to UNITA forces in September 1993, and established a sanctions committee to monitor and report on the implementation of the mandatory

29   Price, T M, 'Article: The Kimberley Process: Conflict Diamonds, WTO Obligations, and the Universality Debate' (2003) 12 *Minnesota Journal of Global Trade* 1, 8–11; Global Witness, 'A Rough Trade: The Role of Companies and Governments in the Angolan Conflict' (Report, Global Witness Ltd, 1998) 7.
30   Price, T M, 'Article: The Kimberley Process: Conflict Diamonds, WTO Obligations, and the Universality Debate' (2003) 12 *Minnesota Journal of Global Trade* 1, 11.
31   Report of the Panel of Experts on Violations of Security Council Sanctions Against UNITA, UN Doc S/2000/203 (10 March 2000) 25.
32   Price, T M, 'Article: The Kimberley Process: Conflict Diamonds, WTO Obligations, and the Universality Debate' (2003) 12 *Minnesota Journal of Global Trade* 1, 9–10.
33   Report of the Panel of Experts on Violations of Security Council Sanctions Against UNITA, UN Doc S/2000/203 (10 March 2000) 25–26.
34   Juma, L, 'The War in Congo: Transitional Conflict Networks and the Failure of Internationalism' (2006) 10 *Gonzaga Journal of International Law* 97, 135–136.
35   Report of the Panel of Experts on Violations of Security Council Sanctions Against UNITA, UN Doc S/2000/203 (10 March 2000) 25, 32.

measures.[36] Despite the agreement of both parties to the Lusaka Peace Accord in November 1994, three years later it was clear that UNITA had used the period of peace to make extensive military preparations funded by its diamond mining activities.[37] In 1997 there was a global diamond recession, which affected the nature of the Angolan conflict. At this time, UNITA withdrew from Cuango Valley mines, cutting back supplies in an overstocked industry. UNITA attempted to close down Angola's official diamond industry by attacking government mining projects. As a result, it was difficult for the government to gain any profit from diamond resources.[38] The Security Council responded with increased pressure on senior UNITA leaders and their immediate families, prohibiting their access to transportation or transit through the territory of other countries.[39]

UNITA have been key players in Angolan diamond production and in the international diamond business since the late 1980s. They have retained a predominant but shifting control over many of the major diamond areas, such as the Cuango River valley and the Lundas, both important areas of production. Between 1992 and 1994, UNITA controlled 90 per cent of Angolan diamond exports. In 1995, UNITA lost control of many areas and its percentage of exports changed. During 1996 and 1997, UNITA was producing about two-thirds of all diamonds mined in Angola. During 1998, the return of former UNITA areas to state administration took place, a condition of the 1994 Lusaka Protocol. UNITA's withdrawal from key areas, such as the

---

36   SC Res 864, UN SCOR, 3277th mtg, S/RES/864 (5 June 1998) (Angola) [19]–[23]; Price, T M, 'Article: The Kimberley Process: Conflict Diamonds, WTO Obligations, and the Universality Debate' (2003) 12 *Minnesota Journal of Global Trade* 1, 10.

37   Lusaka Protocol, UN SCOR, 17th sess, UN Doc S/1994/1441 (22 December 1994); Progress Report of the Secretary-General on the United Nations Observer Mission in Angola (MONUA), UN Doc S/1997/640 (13 August 1997); Price, T M, 'Article: The Kimberley Process: Conflict Diamonds, WTO Obligations, and the Universality Debate' (2003) 12 *Minnesota Journal of Global Trade* 1, 10.

38   Global Witness, 'A Rough Trade: The Role of Companies and Governments in the Angolan Conflict' (Report, Global Witness Ltd, 1998) 1, 7, 12; Maggi, M, 'The Currency of Terrorism: An Alternative Way to Combat Terrorism and End the Trade of Conflict Diamonds' (2003) 15 *Pace International Law Review* 513, 525.

39   SC Res 1127, 3814th mtg, UN Doc S/RES/1127 (28 August 1997) (Angola); Price, T M, 'Article: The Kimberley Process: Conflict Diamonds, WTO Obligations, and the Universality Debate' (2003) 12 *Minnesota Journal of Global Trade* 1, 8–11; Kaplan, M, 'Junior Fellows' Note: Carats and Sticks: Pursuing War and Peace through the Diamond Trade' (2003) 35 *New York University Journal of International Law and Politics* 559, 574–575.

lower Cuango Valley, had a major impact on its level of production, with revenue estimated to be US$200 million for 1998; a major decline from previous years.[40]

Between 1992 and 1998, UNITA obtained an estimated minimum revenue of US$3.72 billion in diamond sales — not including revenue from other sources, or interest generated in overseas bank accounts.[41] By this time, the international community had begun to recognise the critical link between the international diamond trade and UNITA's financial viability. In particular, De Beers was embarrassed by a Global Witness report, which focused on De Beers' annual reporting in relation to their Angolan trading policies. De Beers' annual reports indicated that its policy of buying out most of the rough diamonds on the market had continued, even when it was clear that Angolan diamonds in the 1990s were mined almost entirely by UNITA. It should be noted that in the wake of media criticism, De Beers announced an embargo on the purchase of all diamonds originating from Angola in October 1999 and went on to be a key supporter of the Kimberley Process Certification Scheme.[42]

Further pressure was applied by the UNSC, which adopted resolutions 1173 and 1176 in 1998, prohibiting the direct or indirect export of unofficial Angolan diamonds — those diamonds not accompanied by a government-issued certificate of origin.[43] However, United Nations reports allege that a number of countries acted as intermediaries between UNITA and Antwerp-based diamond traders, including Burkina Faso, Namibia, South Africa, and Zambia, and that Antwerp largely turned a blind eye to the conflict diamonds traffic passing through its diamond market.[44] For example, the government of Belgium reported that Zambian diamond exports to Belgium between February and May 2001 totalled 35,614.14 carats, with an estimated value

---

40   Global Witness, 'A Rough Trade: The Role of Companies and Governments in the Angolan Conflict' (Report, Global Witness Ltd, 1998) 4. Its estimates were made using statistics from World Diamond Industry, Directory and Yearbook 1996/97, Diamond International 1997 and EIU, Country Profile 1997–1998.

41   Ibid 4.

42   Ibid 6–8; Koyame, M, 'United Nations Resolutions and the Struggle to Curb the Illicit Trade in Conflict Diamonds in Sub-Saharan Africa' (2005) 1 *African Journal of Legal Studies* 80, 85.

43   Price, T M, 'Article: The Kimberley Process: Conflict Diamonds, WTO Obligations, and the Universality Debate' (2003) 12 *Minnesota Journal of Global Trade* 1, 8–11.

44   Report of the Panel of Experts on Violations of Security Council Sanctions Against UNITA, UN Doc S/2000/203 (10 March 2000) 27–29.

of $13.3 million, which is 20 times the officially recorded Zambian diamond exports between 1995 to 1998 at $564,272. In addition, diamonds exported by Zambia between 1998 and 2001 had an average carat value of $373.45, indicating that they were more likely to be high-quality gems of Angolan origin than Zambian diamonds.[45]

In 1999, the government captured the crucial UNITA strongholds of Andulo and Bailundo, and forced Savimbi into exile. The offensive cost UNITA its diamond-mining areas, although UNITA profited for some time from stockpiles it had already created. UNITA remained connected to the international diamond markets by air shipping through third countries such as Zambia. In 2002, Savimbi was killed, the Angolan Government and UNITA called a ceasefire, and UNITA became a political party under new leader Samakuva.[46]

Since the establishment of the Kimberley Process in 2002, Angola has played an active part, particularly in the Working Group on Artisanal Mining. It is ironic, in the light of its KP participation, that, unfortunately, the human rights situation in Angola's artisanal fields, which border the DRC, deteriorated dramatically in 2003. In 2003, Angola began a policy of expelling Congolese artisanal miners from Angolan diamond fields. In 2003 and 2004, tens of thousands of Congolese miners were expelled by the Angolan military, creating a refugee crisis in the neighbouring DRC. The first major waves of some 25,000 illegal Congolese miners were expelled in 2003, followed by another 10,000 in February 2004. In April, the UN Office for the Coordination of Humanitarian Affairs (OCHA) reported the arrival of 68,000 exhausted Congolese in the DRC border provinces of Bandundu, Kasai Occidentale, Kasai Orientale, and Katanga. Estimates suggest that approximately 100,000 illicit miners had been expelled from Angola by mid-2004, about a third of the estimated number of miners in Angola.[47]

---

45 Supplementary Report of the Monitoring Mechanism on Sanctions Against UNITA, UN Doc 2001/966 (12 October 2001) 141; Koyame, M, 'United Nations Resolutions and the Struggle to Curb the Illicit Trade in Conflict Diamonds in Sub-Saharan Africa' (2005) 1 *African Journal of Legal Studies* 80, 85–86.

46 Kaplan, M, 'Junior Fellows' Note: Carats and Sticks: Pursuing War and Peace through the Diamond Trade' (2003) 35 *New York University Journal of International Law and Politics* 559, 573–578.

47 Partnership Africa Canada, 'Diamond Industry Annual Review: Republic of Angola 2004' (Annual Report, July 2004) 8–9.

While the expulsions were occurring, UN agencies, Human Rights Watch, and Médecins sans Frontières publicised concerns about abuses reported by returning miners, including rape, body cavity searches of both sexes for hidden diamonds, and general brutality. A human rights group, Voix des Sans Voix, reported that Angolan troops and civilians had subjected many of the Congolese to beatings and death threats.[48]

Protests from the government of the DRC led to an agreement between the two countries that expulsions would be handled in a more co-ordinated and less repressive manner. Although the government of Angola made it plain that the expulsions would continue, Angola acknowledged the military brutality. 'These excesses provoked harmful repercussions, which we regret, and for which we offer a public apology', said Angola's Interior Minister Osvaldo Serra Van-Dúnem. One of the repercussions was a desperate food shortage among returning Angolan refugees in Malanje Province, unable to access markets just across the border in the DRC. In June 2004, the World Food Programme said the Angolan Government's forced repatriation of Congolese nationals had caused hostility towards Angolans who depended on neighbouring Congolese markets to purchase food and other necessary items. An estimated 17,000 Angolans were affected.[49]

Unfortunately, in subsequent years, cross-border expulsions by Angola and attendant human rights abuses continued. As recently as November 2010, UNICEF reported that more than 650 women and girls had been raped during mass expulsions. Approximately 6,621 Congolese returnees arrived in the Western Kasai province in the DRC in two waves in September and October 2010. The reports of sexual violence are based on evidence collected by NGO welcome committees in the region. Many of the victims reported being locked up in derelict buildings, gang-raped and tortured by Angolan security forces, and then forced to walk several days back across the border into the DRC.[50]

The scale of the mass deportations, involving systematic rape and abuse, suggest that they meet the indicia of crimes against humanity attracting the jurisdiction of the ICC. Unfortunately, the issue has not apparently attracted significant attention by either the ICC or the

---

48   Ibid.
49   Ibid.
50   Davies, C, *Reciprocal Violence: Mass Expulsions Between Angola and DRC* (17 February 2011) The Human Rights Brief. Available at: hrbrief.org/2011/02/reciprocal-violence-mass-expulsions-between-angola-and-the-drc/.

Kimberley Process itself. Unlike the situation in Zimbabwe, there has been relatively less media discussion of the Angola deportations. It would appear, on its face, that the connection between artisanal mining in Angola and the commission of these international crimes would qualify Angolan diamonds as being conflict diamonds.

A case study, based on an interview with 28-year-old Dallas Kabungo, is illustrative of the experience of many thousands of DRC citizens expelled from Angola:

> The road north from the Congolese border town of Kamoko barely merits the name; a narrow rutted track, impassable save by toughest 4x4, clogged in early June this year by a tired stream of people flooding north from Angola. It was here that the *Annual Review* encountered 28-year-old Dallas Kabungo.
>
> He had no money, few clothes, and nothing but flip-flops on his feet. He had no idea where to find his wife and child. He'd been walking that road, and others like it on the Angolan side of the border, for over five days, since the night the Angolan police and army surrounded his encampment at Tchiamba, near the town of Lucapa in Lunda Norte. They began by firing shots in the air. Everyone was rounded up, and those without Angolan papers were searched down to their underwear. Anything of value was confiscated. Kabungo lost his spare clothes, a radio, and US$600. 'You came to this country with nothing,' the soldiers told him, 'you will leave with nothing.' Those who resisted the search were beaten, or whipped with belts …
>
> Meanwhile, after waiting a week in the Congolese border town of Kamoko, Dallas Kabungo was finally re-united with his wife, Chantal, and their three-year-old daughter. Soldiers had arrived at the house in Lucapa that Kabungo had bought for her with his diamond earnings. They looted the furniture, took her radio and money, and set her on the road north. It had taken her days of walking in the heat and dust to reach the border.
>
> Their reunion was a bittersweet affair. Kabungo learned that his wife, coming over the border crossing at Myanda, had been raped repeatedly by Angolan border guards. Among his Baluba tribe, he said, it's believed that if a woman engages in adultery, her children soon fall sick and die. He's not sure if the curse works when the woman has been raped.[51]

---

51   Partnership Africa Canada, 'Diamonds and Human Security: Annual Review 2008' (Annual Report, October 2008) 14.

Grave human rights violations by the Angolan security forces, namely the Angolan military and national police, have also targeted Angolan citizens engaged in artisanal mining. One documented case, the killing of Belito Mendes, occurred on Saturday 12 May 2007. The victim, 28-year-old Belito Mendes, a veteran of the Angolan army, was beaten to death by members of the Angolan National Police after refusing to hand over the small amount of money he had on his person.[52] It is unfortunate that the situation in Angola received no attention by the Kimberly Process at the time that these human rights violations were occurring.

While the Angolan expulsions have attracted little attention, there have recently been precedent-setting national prosecutions in French courts, in relation to arms trafficking and bribery related to the Angolan civil war of the 1990s. On 28 October 2009, a Paris court convicted 36 people in connection to illegal arms sales to Angola during its civil war, including arms dealers, middlemen, and French politicians. Arms trading went hand in hand with diamond and oil trading during the war, as sales of natural resources were used to purchase armaments, which were then turned on the civilian population. Arms dealers Pierre Falcone and Arkadi Gaydamak were sentenced to six years in prison for arms trafficking and other offences. Former French Interior Minister Charles Pasqua was sentenced to a year in jail for taking bribes from the two men. Amongst others convicted, a son of former president François Mitterrand and a banker from BNP Paribas, a top French bank, were given suspended sentences. At the time of reporting, appeals were expected to follow.[53]

While the problem of international human rights crimes is front and centre when considering the rough diamond industry in Angola, lower level human rights issues present further challenges. In particular, the Angolan artisanal industry has a significant child labour problem. According to UNICEF, 70 per cent of Angola's population is under 24, and 30 per cent of children between the ages of five and 14 years work. A 2004 case study undertaken by Partnership Africa Canada and Global Witness in the Lunda Norte province showed that family

---

52   Ibid 13–14.
53   Global Witness, *Global Witness Welcomes French 'Angolagate' Verdict as Victory for Justice* (28 October 2009). Available at: www.globalwitness.org/en/archive/global-witness-welcomes-french-angolagate-verdict-victory-justice/.

mining groups consist of women and children as well as adult men. Forty-six per cent of those interviewed and working were children in the age group 5–16. Many women worked as well, and differences in gender representation were large in only one age group — young men dominated the 17–25 age group. The report noted:

> In today's mining areas, fear, insecurity and sexual abuse are constant. Today's child miners are ... a direct result of war, poverty and the absence of education; there are few schools in the diamond regions and even the existing ones were destroyed during the many decades of war.

Yet these families, who worked in unsafe and abusive conditions, derived less than 5 per cent of their income from mining diamonds, with the largest part derived from agriculture and the rest from business and trading. This was not because diamonds represented less work than agriculture; it was because the diggers were so badly paid for the diamonds they found.[54]

Sierra Leone

The conflict in Sierra Leone saw the problem of conflict diamonds reach a greater level of notoriety, through the activities of the RUF. It was through the activities of the RUF that the illegal diamonds trade became connected in the eyes of the international community not only with the prolongation of conflict, but also with the perpetration of graphic human rights violations. The terror tactics employed by the RUF to subdue the local civilian population included the amputation of hands, limbs, and body parts. The militia also perpetrated unlawful killings, physical and sexual violence against civilians, abductions, looting and destruction of civilian property, forced labour including sexual slavery, and the conscription of boys and girls into the armed forces and their deployment in active fighting. The names of their military operations, 'Operation Pay Yourself' and 'Operation No Living Thing', were illustrative of their intentions and encouraged

---

54   Partnership Africa Canada, 'Diamond Industry Annual Review: Republic of Angola 2005' (Annual Report, July 2005) 9; Global Witness and Partnership Africa Canada, 'Rich Man, Poor Man, Development Diamonds and Poverty Diamonds: The Potential for Change in the Artisanal Alluvial Diamond Fields of Africa' (Report, October 2004) 17–23.

the commission of these crimes.[55] The conflict resulted in the loss of 75,000 lives, created half a million refugees and internally displaced 2.25 million, while an estimated 12,000 children were abducted to fight as soldiers. Through the infamous practice of amputation, some 20,000 of the civilian population were left mutilated. In the quintessential paradox of the resource curse, Sierra Leone was listed at the bottom of the United Nations Development Programme (UNDP) Human Development Index in 2001, despite its abundance of natural resources.[56]

In 1991, former army corporal Foday Sankoh emerged as the leader of the RUF, which attacked Sierra Leonean border towns from Liberia. The attacks were marked by brutality against civilians, and children were kidnapped and inducted into the RUF.[57] In 1994, the RUF overran diamond-rich areas, bauxite and titanium mines, thereby bankrupting the economy, but providing themselves with access to an abundance of natural resources. A peace accord was signed in 1996 by newly elected President Kabbah and Foday Sankoh, but soldiers seized power the following year under the banner of the Armed Forces Revolutionary Council (AFRC). Major Johnny Paul Koroma became the chairman and invited the RUF to join the government, resulting in systematic human rights abuses and the destruction of the formal economy.[58]

The UNSC responded by imposing an arms embargo on Sierra Leone in 1997. The United Nations launched a limited peacekeeping operation, United Nations Observer Mission Sierra Leone (UNOMSIL), consisting of 70 observers. However, it was the regional peacekeeping force Economic Community of West African States Cease-Fire Monitoring Group (ECOMOG) that made the decisive intervention. In February 1998, 10,000 to 12,000 ECOMOG troops forced the RUF/AFRC out of

---

55   Transcript of Proceedings, 'Opening Statement of the Prosecution', *The Prosecutor v Issa Hassan Sesay, Morris Kallon and Augustine Gbao (Trial)*, (Trial Chamber I, Special Court for Sierra Leone, Case No SCSL-04-15-T, 5 July 2004) 26; Crane, D M, 'Terrorists, Warlords and Thugs' (2006) 21 *American University International Law Review* 505, 515–516.

56   Price, T M, 'Article: The Kimberley Process: Conflict Diamonds, WTO Obligations, and the Universality Debate' (2003) 12 *Minnesota Journal of Global Trade* 1, 13; Koyame, M, 'United Nations Resolutions and the Struggle to Curb the Illicit Trade in Conflict Diamonds in Sub-Saharan Africa' (2005) 1 *African Journal of Legal Studies* 80, 86.

57   Transcript of Proceedings, 'Opening Statement of the Prosecution', *The Prosecutor v Issa Hassan Sesay, Morris Kallon and Augustine Gbao (Trial)*, (Trial Chamber I, Special Court for Sierra Leone, Case No SCSL-04-15-T, 5 July 2004) 20.

58   Smillie, I, L Gberie and R Hazleton, 'The Heart of the Matter: Sierra Leone, Diamonds and Human Security' (Report, Partnership Africa Canada, 2000) 3.

the capital city, Freetown, and engaged their forces in the countryside, enabling the Kabbah Government to be re-established. During the same year, an embargo was imposed that allowed the government to rearm itself, but the embargo on RUF weapons was maintained.[59]

The RUF, however, outmanoeuvred the embargo, and their offensive in January 1999 resulted in the capture of Freetown. In a period of only two weeks, Freetown witnessed the torture and murder of cabinet ministers, journalists, and civil servants, the deaths of some 6,000 civilians, and the disappearance of 2,000 children. The RUF were ultimately pushed out of Freetown by ECOMOG forces. In an effort to protect Freetown, maintain security, and train Sierra Leone's army, UN peacekeepers were increased to approximately 13,000, augmented by another 750 from the United Kingdom. A serious challenge arose with the capture of 500 UN soldiers by the RUF, who were only released when the UN was assisted by British troops.[60]

Peace finally became a possibility following the capture of Sankoh in 2000 and the subsequent disarmament of the RUF. In May 2001, the RUF released children who had been abducted and conscripted — aged 10 to 15 — and in January 2002, the UN announced the completion of the disarmament of over 45,000 rebel soldiers. Kabbah was re-elected president in 2002 by 70 per cent of the vote in the first peaceful election since the civil war.[61] Following certificate of origin initiatives such as the Kimberley Process, the legitimate diamond industry in Sierra Leone has begun to re-emerge. In 2003, Sierra Leone legally exported $76 million of diamonds from alluvial fields, while the 2004 total was estimated as being $120 million.[62]

59   Ibid 2–3; Price, T M, 'Article: The Kimberley Process: Conflict Diamonds, WTO Obligations, and the Universality Debate' (2003) 12 *Minnesota Journal of Global Trade* 1, 14.
60   Smillie, I, L Gberie and R Hazleton, 'The Heart of the Matter: Sierra Leone, Diamonds and Human Security' (Report, Partnership Africa Canada, 2000) 3; Price, T M, 'Article: The Kimberley Process: Conflict Diamonds, WTO Obligations, and the Universality Debate' (2003) 12 *Minnesota Journal of Global Trade* 1, 14–16.
61   Price, T M, 'Article: The Kimberley Process: Conflict Diamonds, WTO Obligations, and the Universality Debate' (2003) 12 *Minnesota Journal of Global Trade* 1, 14–16.
62   Global Witness and Partnership Africa Canada, 'Rich Man, Poor Man, Development Diamonds and Poverty Diamonds: The Potential for Change in the Artisanal Alluvial Diamond Fields of Africa' (Report, October 2004) 9.

During the civil war, the RUF armed itself through the sale of illegal diamonds, earning about $120 million per year from 1991 to 1999.[63] The alluvial diamond fields in the Kono region and the Tongo Field region were the prize, giving the RUF little reason to engage with the peace process or even to try to win the war. The RUF's mining regime was largely based on forced labour, whereby civilians, including children, were tied up, forbidden to speak, and forced to work 12-hour shifts at gunpoint. They were not paid or fed, and sustained themselves by eating nearby fruit. New labour was brought in when existing workers became too sick to work or were shot. Shootings were carried out by the RUF small boys' units, staffed by children as young as 11 years old, who were armed with AK-47s.[64] Expert evidence suggests that the reason that the RUF engaged in amputations and fear tactics was to maintain complete control over the diamond mining fields without interference by the thousands of freelance diggers who would otherwise also mine the area.[65] Prior to the RUF invasion, the diamond fields had been mined by thousands of licensed and unlicensed artisanal diggers, policed by a government force numbering 500 persons with access to helicopters.[66]

From the very inception of the civil war, the RUF allegedly received the support of Liberian President Charles Taylor. He encouraged Foday Sankoh to mine diamonds and gold from the Kono district to finance the war, and trade these goods with Burkina Faso and Libya

---

63   Report of the Panel of Experts Appointed Pursuant to Security Council Resolution 1306 (2000), paragraph 19, in Relation to Sierra Leone, UN Doc S/2000/1195 (20 December 2000) 578–579; Price, T M, 'Article: The Kimberley Process: Conflict Diamonds, WTO Obligations, and the Universality Debate' (2003) 12 *Minnesota Journal of Global Trade* 1, 14.

64   Transcript of Proceedings, 'Opening Statement of the Prosecution', *The Prosecutor v Issa Hassan Sesay, Morris Kallon and Augustine Gbao (Trial)*, (Trial Chamber I, Special Court for Sierra Leone, Case No SCSL-04-15-T, 5 July 2004) 27; Mitchell III, A F, 'Sierra Leone: The Road to Childhood Ruination Through Forced Recruitment of Child Soldiers and the World's Failure to Act' (2003) 2 *Regent Journal of International Law* 81, 86–87; Woody, K E, 'Diamonds on the Soul of Her Shoes: The Kimberley Process and the Morality Exception to WTO Restrictions' (2007) 22 *Connecticut Journal of International Law* 335, 335.

65   Transcript of Proceedings, 'Evidence of Expert Witness Ian Smillie', *The Prosecutor v Charles Taylor (Trial)*, (Special Court for Sierra Leone, Trial Chamber II, Case No SCSL-03-01-PT, 7 January 2008) 581–582.

66   Kaplan, M, 'Junior Fellows' Note: Carats and Sticks: Pursuing War and Peace through the Diamond Trade' (2003) 35 *New York University Journal of International Law and Politics* 559, 568–570; Transcript of Proceedings, 'Evidence of Expert Witness Ian Smillie', *The Prosecutor v Charles Taylor (Trial)*, (Special Court for Sierra Leone, Trial Chamber II, Case No SCSL-03-01-PT, 7 January 2008) 503–505, 533.

for supplies, ammunition, and weapons.[67] In return for diamonds, Taylor allegedly supplied the RUF with consignments of AK-47s, RPGs, Uzis, and ammunition, and provided military training to the militia.[68] A UN expert panel collected evidence of cargoes of weapons in the period 1998 to 1999 being airlifted from the Ukraine and Eastern Europe to Liberia via transit stops in Burkina Faso and Niger. The weapons were then diverted to the RUF, who made use of them in their offensive against Freetown in January 1999 and other operations. It is interesting to note that one of the individuals cited as a key arms trader, Russian national Viktor Bout, has been tried and convicted in the United States on charges related to illegal arms dealing.[69]

Even prior to the imposition of diamond trading sanctions on Sierra Leone, it became apparent that Sierra Leonean diamonds were reaching the international market after first being diverted through other countries including, most notably, Liberia. Liberian annual diamond production capacity is estimated as being 100,000 to 150,000 carats, however, rough diamond exports to Antwerp from Liberia were recorded at 31 million carats from 1994 to 1998, an average of six million carats per year. Similarly, Côte d'Ivoire, which had not been mining diamonds since the mid-1980s, registered exports of more than 1.5 million between 1995 and 1997. Guinea also appeared to be

---

67   Transcript of Proceedings, 'Opening Statement of the Prosecution', *The Prosecutor v Issa Hassan Sesay, Morris Kallon and Augustine Gbao (Trial)*, (Trial Chamber I, Special Court for Sierra Leone, Case No SCSL-04-15-T, 5 July 2004) 21.

68   Ibid 48; Torbey, C, 'The Most Egregious Arms Broker: Prosecuting Arms Embargo Violators in the International Criminal Court' (2007) 25 *Wisconsin International Law Journal* 335, 338; Woodward, L, 'Taylor's Liberia and the UN's Involvement' (2003) 19 *New York Law School Journal of Human Rights* 923, 932–933.

69   Transcript of Proceedings, 'Evidence of Expert Witness Ian Smillie', *The Prosecutor v Charles Taylor (Trial)*, (Special Court for Sierra Leone, Trial Chamber II, Case No SCSL-03-01-PT, 7 January 2008) 550–562; Report of the Panel of Experts Appointed Pursuant to Security Council Resolution 1306 (2000), paragraph 19, in Relation to Sierra Leone, UN Doc S/2000/1195 (20 December 2000) 34–36, 41; Stempel, J, 'Russian Arms Dealer Viktor Bout's U.S. Conviction Upheld', Reuters, 27 September 2013. Available at: www.reuters.com/article/us-usa-crime-bout-idUSBRE98Q0PG20130927.

a diversion point for Sierra Leonean diamonds.[70] In 2000, following the passing of United Nations Security Council Resolution 1306, banning the sale of diamonds from Sierra Leone, Liberian production increased 161 per cent from 1999 levels. In response to a UN panel of experts report, Security Council Resolution 1343 was passed, prohibiting the export or import of Liberian diamonds, so as to close the Liberian way-station for Sierra Leonean diamonds.[71]

According to Global Witness, the RUF trade in conflict diamonds also had connections to Al Qaeda and the world of international terrorism. For example, in 1998 an RUF official met with operatives from Al Qaeda to sell diamonds, with the first transaction involving $100,000 in cash for a parcel of diamonds. As further evidence of the connection, in 2004 Ahmed Khalfan Ghailani, a high-level Al Qaeda operative from

---

70   Kaplan, M, 'Junior Fellows' Note: Carats and Sticks: Pursuing War and Peace through the Diamond Trade' (2003) 35 *New York University Journal of International Law and Politics* 559, 568–570; Smillie, I, L Gberie and R Hazleton, 'The Heart of the Matter: Sierra Leone, Diamonds and Human Security' (Report, Partnership Africa Canada, 2000) 6; Koyame, M, 'United Nations Resolutions and the Struggle to Curb the Illicit Trade in Conflict Diamonds in Sub-Saharan Africa' (2005) 1 *African Journal of Legal Studies* 80, 87; Cuellar, M, 'Panel: Combating Diamonds in Sub-Saharan Africa' (2007) 1 *African Journal of Legal Studies* 80, 447–448. The literature also notes that identifying accurately even the 'country of export' was complicated by problems of fraud and corruption within the Antwerp diamond market at the time. Smillie, I, L Gberie and R Hazleton, 'The Heart of the Matter: Sierra Leone, Diamonds and Human Security' (Report, Partnership Africa Canada, 2000) 6. It was also noted that Liberia was used as a way-station for other illicit diamonds — Russian diamonds, for example. Transcript of Proceedings, 'Evidence of Expert Witness Ian Smillie', *The Prosecutor v Charles Taylor (Trial)*, (Special Court for Sierra Leone, Trial Chamber II, Case No SCSL-03-01-PT, 7 January 2008) 565–567.

71   Kaplan, M, 'Junior Fellows' Note: Carats and Sticks: Pursuing War and Peace through the Diamond Trade' (2003) 35 *New York University Journal of International Law and Politics* 559, 570. Resolutions renewed the sanctions on a regular basis, even following the election of the reformist government of President Sirleaf on 23 November 2005, and the arrest of former President Charles Taylor on war crimes charges on 29 March 2006. However, trade in diamonds possessing a certificate of origin was permitted. Controversially, the sanctions were entirely removed in 2007 under SC Res 1753 of 27 April 2007, with commentators arguing that Liberia was still not in a position to ensure the conflict-free status of its diamonds. Global Witness, 'Cautiously Optimistic: The Case for Maintaining Sanctions in Liberia' (Report, 2006) 17.

Tanzania was arrested on suspicion of being involved in the trading of conflict diamonds and running a $20 million financing operation, trading illegal conflict diamonds in the Liberian capital of Monrovia.[72]

Crimes fuelled by conflict diamonds have now been prosecuted in a number of cases before the international criminal justice system, namely the Special Court for Sierra Leone. After Sankoh was captured in 2000, the Special Court was established to try him and other persons for violations of international law during the conflict. Although Sankoh died before the commencement of his trial in 2004, other members of the RUF leadership, namely Issa Hassan Sesay, Morris Kallon, and Augustine Gbao, were convicted of international crimes on 2 March 2009, with their convictions upheld on appeal on 26 October 2009.[73] In June 2003, the Sierra Leone Special Court unsealed an indictment against Liberian President Charles Taylor in relation to his alleged involvement in violations of international criminal law during the Sierra Leone conflict. President Taylor subsequently resigned his office and went into hiding in Nigeria in August 2003, until being taken into custody by the court on 29 March 2006.[74] His trial, which was moved to The Hague, commenced on 7 January 2008, with closing arguments concluding on 11 March 2011. Charles Taylor was found

---

72   Martinez, I, 'Africa at the Crossroads: Current Themes in African Law: VI. Conflict Resolution in Africa: Sierra Leone's "Conflict Diamonds": The Legacy of Imperial Mining Laws and Policy' (2001/2002) 10 *University of Miami International and Comparative Law Review* 217, 236; Global Witness, 'For a Few Dollars More: How Al Qaeda Moved into the Diamond Trade' (Report, April 2003) 45–50; Tanna, Ketan, 'Pakistan Arrests Conflict Diamonds Al Qaeda Operative', *The Rapaport Diamond Report* (2 August 2004), cited in Koyame, M, 'United Nations Resolutions and the Struggle to Curb the Illicit Trade in Conflict Diamonds in Sub-Saharan Africa' (2005) 1 *African Journal of Legal Studies* 80, 95. For discussion of the role of ECOMOG in such interventions, see Brockman, J, 'Liberia: The Case for Changing UN Processes for Humanitarian Interventions' (2004) 22 *Wisconsin International Law Journal* 711, 713–720.

73   Transcript of Proceedings, 'Opening Statement of the Prosecution', *The Prosecutor v Issa Hassan Sesay, Morris Kallon and Augustine Gbao (Trial)*, (Trial Chamber I, Special Court for Sierra Leone, Case No SCSL-04-15-T, 5 July 2004); *The Prosecutor v Issa Hassan Sesay, Morris Kallon and Augustine Gbao (Trial Judgement)* (Special Court for Sierra Leone, Trial Chamber I, Case No SCSL-04-15-T, 2 March 2009); *The Prosecutor v Issa Hassan Sesay, Morris Kallon and Augustine Gbao (Appeal Judgement)* (Special Court for Sierra Leone, Appeals Chamber, 29 October 2009). For a comprehensive analysis of the early work of the Tribunal, see Jalloh, C C, 'The Contribution of the Special Court for Sierra Leone to the Development of International Law' (2007) 15 *RADIC* 165, 165–207. See also Schocken, C, 'The Special Court for Sierra Leone: Overview and Recommendations' (2002) 20 *Berkeley Journal of International Law* 436.

74   Brockman, J, 'Liberia: The Case for Changing UN Processes for Humanitarian Interventions' (2004) 22 *Wisconsin International Law Journal* 711, 738–739; Special Court for Sierra Leone, Office of the Prosecutor, 'Chief Prosecutor Announces the Arrival of Charles Taylor at the Special Court' (Press Release, 29 March 2006). Available at: www.rscsl.org/Documents/Press/OTP/prosecutor-032906.pdf.

guilty on 26 April 2012 on all 11 counts, when the Trial Chamber delivered its judgement. On 30 May 2012, the Trial Chamber delivered its sentencing judgement, sentencing Taylor to 50 years imprisonment. On 26 September 2013, Taylor's conviction and sentence were upheld by the judgement of the Appeals Chamber.[75] Pro-government forces, most notably the so-called Civil Defence Forces (CDF) militia, have also been brought to account for human rights abuses allegedly committed during the war. For example, the CDF was known to practice torture, while the Sierra Leone Government was also documented as enlisting child soldiers during the military regime of Valentine Strasser from 1992 to 1996.[76] Members of the CDF leadership were convicted of international crimes by the Special Court on 2 August 2007, with the convictions upheld on appeal on 28 May 2008.[77]

Since the conclusion of the Sierra Leone civil war, there has thankfully been a decade of peace and development. It should be noted, however, that issues remain, particularly in relation to Sierra Leone's alluvial diamond mining sector. Of particular concern is the ongoing problem of child labour in the 120,000 person strong artisanal industry.[78]

The Democratic Republic of Congo

The problem of conflict diamonds was central to the outbreak and prolongation of the 1996–2002 war in the DRC, which has now transformed into the recent civil war in the north-east of that country. The DRC has suffered from a resource curse relating not only to diamonds, but a range of minerals including copper, cobalt, uranium, tin, zinc, and coltan, the latter being a key material in the manufacture

---

75   Special Court for Sierra Leone, Residual Special Court for Sierra Leone, *The Prosecutor vs. Charles Gankay Taylor*. Available at: www.rscsl.org/Taylor.html.

76   Pham, J P, 'A Viable Model for International Criminal Justice: The Special Court for Sierra Leone' (2006) 19 *New York International Law Review* 37, 70–72; Mitchell III, A F, 'Sierra Leone: The Road to Childhood Ruination Through Forced Recruitment of Child Soldiers and the World's Failure to Act' (2003) 2 *Regent Journal of International Law* 81, 85–86.

77   *The Prosecutor v Moinina Fofana and Allieu Kondewa (Trial Judgement)* (Special Court for Sierra Leone, Trial Chamber I, Case No SCSL-04-14-T, 2 August 2007); *The Prosecutor v Moinina Fofana and Allieu Kondewa (Appeal Judgement)* (Special Court for Sierra Leone, Appeals Chamber, Case No SCSL-04-14-A, 28 May 2008).

78   Global Witness and Partnership Africa Canada, 'Rich Man, Poor Man, Development Diamonds and Poverty Diamonds: The Potential for Change in the Artisanal Alluvial Diamond Fields of Africa' (Report, October 2004) 9–16.

of mobile telephones and other electronic equipment.[79] The resource curse is seen as working in two different ways in the DRC conflict. Access to resources acted both as a reason to enter the war for militias and national governments, and a factor for prolonging involvement in the war. The war resulted in the deaths of some 3 million people, because of fighting, disease, and malnutrition, as well as approximately 500,000 refugees and 2 million internally displaced people, and saw six national armies from neighbouring countries intervene in the conflict. The war also accentuated the poverty of the population, with the DRC ranked at 155th place out of 173 countries on the UNDP Human Development Index, despite its abundance of natural resources.[80]

The war has its genesis in the 1994 genocide in Rwanda, where militias known as Interahamwe, backed by the government armed forces, killed some 800,000 ethnic Tutsis and moderate Hutus. In the wake of their defeat by the Tutsi-led rebels, the Interahamwe and other *genocidaires* fled across the border into the DRC, from which they launched attacks on Rwanda and the longstanding Tutsi population resident in eastern Congo. In response to these attacks, and in an attempt to protect the Tutsi population, the Alliance of Democratic Forces for the Liberation of Congo (ADFL), led by Laurent Kabila, was formed. The ADFL was backed by the incoming Tutsi-led Rwandan government, which, commentators allege, not only supported the ADFL's wider agenda to topple Congolese President Mobuto Sese Seko, but also had designs on the mineral wealth of the eastern Congolese provinces. In 1997, Kabila's forces entered the capital Kinshasa and toppled Mobutu's government. The 1996 to 1997 phase of the conflict is often termed the First Congolese War.[81]

---

79   Juma, L, 'The War in Congo: Transitional Conflict Networks and the Failure of Internationalism' (2006) 10 *Gonzaga Journal of International Law* 97, 122–124; British Broadcasting Corporation, 'Q&A: Plunder in the Congo', *BBC News: World Edition* (21 October 2002). Available at: news.bbc.co.uk/2/hi/business/2346817.stm.

80   British Broadcasting Corporation, 'Q&A: Plunder in the Congo', *BBC News: World Edition* (21 October 2002). Available at: news.bbc.co.uk/2/hi/business/2346817.stm; Fonseca, A, *Four Million Dead: The Second Congolese War, 1998–2004* (18 April 2004) 49. Available at: www. oocities.org/afonseca/CongoWar.htm. For statistics on casualties, refugees and displacements relating to the 1998–2002 period, see Koyame, M, 'United Nations Resolutions and the Struggle to Curb the Illicit Trade in Conflict Diamonds in Sub-Saharan Africa' (2005) 1 *African Journal of Legal Studies* 80, 89.

81   British Broadcasting Corporation, 'Q&A: Plunder in the Congo', *BBC News: World Edition* (21 October 2002). Available at: news.bbc.co.uk/2/hi/business/2346817.stm; Juma, L, 'The War in Congo: Transitional Conflict Networks and the Failure of Internationalism' (2006) 10 *Gonzaga Journal of International Law* 97, 132–139.

The support of Rwanda and allied Uganda for the new Kabila Government proved to be short-lived. Speculation regarding the change in Rwandan policy ranges from Kabila no longer being interested in protecting Congolese Tutsi, to wider rumours about Kabila reneging on important promises made to large mining corporations.[82] In August 1998 the Congolese Rally for Democracy (RCD) emerged, supported by Rwanda and intent on enforcing a second regime change in Kinshasa.[83] The forward offensive was blocked, however, by troops from Angola, Zimbabwe and Namibia, and a four-year stand-off ensued in what is sometimes called the Second Congolese War.[84]

Although Kabila's government survived the invasion, Kabila himself was not so fortunate, and died to an assassin's bullet in 2001. It was left to his son and successor, Joseph Kabila, to become a party to the September 2002 Luanda Agreement. The agreement saw the commencement of a gradual withdrawal of foreign armies, although conflict has continued through rebel groups operating particularly in the north-eastern Ituri and Kivu regions.[85]

A UNSC expert panel has argued that the involvement of foreign governments and militias in the conflict was influenced by a desire to reap the benefits of the extensive mineral wealth, including diamonds, of the DRC. The panel argued that the main beneficiaries from the conflict have been individuals in the military and political leadership of the intervening nations, rather than those nations themselves.[86]

As evidence of how Rwandan diamond enterprises supported Kabila's ADFL in the First Congolese War, the report detailed a financial transaction where a US$3.5 million payment was made from a diamond company named MIBA to COMIEX, a company

---

82   Juma, L, 'The War in Congo: Transitional Conflict Networks and the Failure of Internationalism' (2006) 10 *Gonzaga Journal of International Law* 97, 144–148.

83   Ibid 145–147.

84   British Broadcasting Corporation, 'Q&A: Plunder in the Congo', *BBC News: World Edition* (21 October 2002). Available at: news.bbc.co.uk/2/hi/business/2346817.stm.

85   Price, T M, 'Article: The Kimberley Process: Conflict Diamonds, WTO Obligations, and the Universality Debate' (2003) 12 *Minnesota Journal of Global Trade* 1, 16–22; British Broadcasting Corporation, 'Q&A: Plunder in the Congo', *BBC News: World Edition* (21 October 2002). Available at: news.bbc.co.uk/2/hi/business/2346817.stm.

86   Report of the Panel of Experts on the Illegal Exploitation of Natural Resources and Other Forms of Wealth of the Democratic Republic of Congo, UN Doc S/2001/357 (12 April 2001) [135]–[138]; Clark, J F and M Koyame, 'The Economic Impact of the Congo War' in J F Clark (ed.), *The Africa Stakes of the Congo War* (Palgrave MacMillan, 2002) 55–56.

owned by Kabila. The payment was from an account held with the Rwandan Government's financial institution Banque de Commerce, du Développement et d'Industrie. The report also stated that cargo flights carrying military equipment to airstrips in eastern DRC would return with loads of gold and coffee, as well as businessmen with stolen diamonds for sale.[87]

During the Second Congolese War, a major diamond and gold dealer named Ali Hussein was reported to have met many times with Rwandan Government officials.[88] Rwanda was estimated to have earned some US$4 million from 1998 to 2000 as a result of diamond licensing revenues, although it later shared the income with the RCD-Goma regional administration.[89] Statistics from the Antwerp Diamond High Council show that Rwanda, which has no diamond resources of its own, exported US$720,000 worth of diamonds in 1997, US$17,000 in 1998, US$439,000 in 1999 and US$1.8 million in 2000.[90] These statistics were largely corroborated by World Trade Organization (WTO) records.[91]

The Ugandan Government funded its presence in the DRC through the re-exportation economy, repackaging and exporting natural resources from the DRC as Ugandan natural resources. According to the Antwerp Diamond High Council, Uganda, which was never previously a diamond exporter, exported $1.5 million dollars worth of diamonds in 1998, $1.8 million in 1999, and $1.3 million in 2000.[92]

---

87   Fonseca, A, *Four Million Dead: The Second Congolese War, 1998–2004* (18 April 2004) 49–50. Available at: www.oocities.org/afonseca/CongoWar.htm; Report of the Panel of Experts on the Illegal Exploitation of Natural Resources and Other Forms of Wealth of the Democratic Republic of Congo, S/2001/357 (12 April 2001) [25]–[29].
88   Report of the Panel of Experts on the Illegal Exploitation of Natural Resources and Other Forms of Wealth of the Democratic Republic of Congo, UN Doc S/2001/357 (12 April 2001) [93]; Fonseca, A, *Four Million Dead: The Second Congolese War, 1998–2004* (18 April 2004) 52. Available at: www.oocities.org/afonseca/CongoWar.htm.
89   Report of the Panel of Experts on the Illegal Exploitation of Natural Resources and Other Forms of Wealth of the Democratic Republic of Congo, UN Doc S/2001/357 (12 April 2001) 29.
90   Ibid 25.
91   Ibid [21]–[30]; Fonseca, A, *Four Million Dead: The Second Congolese War, 1998–2004* (18 April 2004) 54. Available at: www.oocities.org/afonseca/CongoWar.htm.
92   Report of the Panel of Experts on the Illegal Exploitation of Natural Resources and Other Forms of Wealth of the Democratic Republic of Congo, UN Doc S/2001/357 (12 April 2001) 21.

Statistics from the WTO corroborated the fact that Uganda, without diamond resources of its own, exported diamonds between 1998 and 2001.[93]

Zimbabwe was also benefiting from Congolese diamonds, particularly in the diamond-rich Mbuji-Mayi area. There were notably 2,500 to 3,000 Zimbabwean troops in the area, even after most other foreign armies had withdrawn from the DRC in 2002.[94] The region had previously been the site of a major battle between Rwandan-backed forces and DRC forces in 2001.[95] Burundi also caused Kabila consternation when its military venture into South Kivu province appeared to threaten the Congolese diamond reserves in nearby Katanga.[96] WTO statistics showed that Burundi, without any diamond extraction industries of its own, exported sizeable quantities of diamonds in the period 1997–2000.[97] In an interesting twist, commentators suggest that Angola's involvement was largely concerned with the disruption of the conflict diamonds trade rather than its perpetuation, considering that Kabila was opposed to UNITA forces trading their diamonds on Congolese territory.[98]

Mineral resources were also a goal for a variety of militia groups, some of which were allegedly proxies for foreign governments. As a result, diamond-rich areas became the focus of fighting during the conflict. In August 1999, the diamond-rich town of Kisangani was the site of a major battle following the split of the main rebel group, which was fuelled by the desire to control local natural resources. The RCD, split into two parts: the RCD-ML, supported by Uganda, and the RCD-Goma, supported by Rwanda. The battle, which also involved direct fighting between Ugandan and Rwandan troops, killed some

---

93   Ibid [21]–[30]; Fonseca, A, *Four Million Dead: The Second Congolese War, 1998–2004* (18 April 2004) 54. Available at: www.oocities.org/afonseca/CongoWar.htm.

94   Juma, L, 'Africa, its Conflicts and its Traditions: Debating a Suitable Role for Tradition in African Peace Initiatives' (2005) 13 *Michigan State University College of Law* 417, 454; Juma, 'The War in Congo', above n 69, 148–149; BBC, 'Plunder in the Congo', above n 115.

95   Report of the Panel of Experts on the Illegal Exploitation of Natural Resources and Other Forms of Wealth of the Democratic Republic of Congo, UN Doc S/2001/357 (12 April 2001) 36.

96   Fonseca, A, *Four Million Dead: The Second Congolese War, 1998–2004* (18 April 2004) 31. Available at: www.oocities.org/afonseca/CongoWar.htm.

97   Ibid 54; Report of the Panel of Experts on the Illegal Exploitation of Natural Resources and Other Forms of Wealth of the Democratic Republic of Congo, UN Doc S/2001/357 (12 April 2001) [21]–[30].

98   Juma, L, 'The War in Congo: Transitional Conflict Networks and the Failure of Internationalism' (2006) 10 *Gonzaga Journal of International Law* 97, 149.

400 soldiers and 200 civilians.[99] On 18 November 2000, RCD-Goma and its Rwandan allied troops stationed in Kisangani attacked and took control of positions belonging to Ugandan-supported forces in Bengamisa, 50 kilometres north-west of Kisangani, an area with an abundance of diamonds.[100] Late in December 2000, the RCD-Goma launched attacks against two other diamond-rich areas in Kandole and Lakutu.[101] Control over diamond resources caused a further schism when the RCD-National formed with the objective of controlling north-eastern diamond resources in the Ituri region.[102]

The RCD-ML benefited from its control over resources in the Ituri region during the Second Congolese War. For example, RCD-ML received benefits in exchange for granting the Ugandan-Thai company DARA-Forest a licence to harvest diamonds, coltan, and timber after the company had failed to acquire a concession from Kinshasa in March 1998.[103] The expert panel identified the retired Ugandan Major General Salim Saleh as benefiting from Congolese diamonds, particularly from the Kisangani region, as well as gold. He allegedly played a major role in Uganda's operations in north-eastern Congo, cultivating a reciprocal relationship with RCD-ML elites. In essence, he ensured their individual safety in return for their looking after his resource extraction schemes. The general also fostered a close relationship with the senior leadership of RCD-National, through which dealings in diamonds and other natural resources were facilitated.[104]

In the wake of the phased withdrawal of foreign troops following the 2002 Luanda Agreement, the UNSC expert panel commented that governments, including DRC government members, have continued

99 Fonseca, A, *Four Million Dead: The Second Congolese War, 1998–2004* (18 April 2004) 32, 25. Available at: www.oocities.org/afonseca/CongoWar.htm; Juma, L, 'Africa, its Conflicts and its Traditions: Debating a Suitable Role for Tradition in African Peace Initiatives' (2005) 13 *Michigan State Journal of International Law* 417, 454.
100 Report of the Panel of Experts on the Illegal Exploitation of Natural Resources and Other Forms of Wealth of the Democratic Republic of Congo, UN Doc S/2001/357 (12 April 2001) 37.
101 Ibid.
102 Fonseca, A, *Four Million Dead: The Second Congolese War, 1998–2004* (18 April 2004) 42. Available at: www.oocities.org/afonseca/CongoWar.htm.
103 Ibid 51; Report of the Panel of Experts on the Illegal Exploitation of Natural Resources and Other Forms of Wealth of the Democratic Republic of Congo, UN Doc S/2001/357 (12 April 2001) [48]–[49].
104 Report of the Panel of Experts on the Illegal Exploitation of Natural Resources and Other Forms of Wealth of the Democratic Republic of Congo, UN Doc S/2001/357 (12 April 2001) [88]–[89]; Fonseca, A, *Four Million Dead: The Second Congolese War, 1998–2004* (18 April 2004) 52–53. Available at: www.oocities.org/afonseca/CongoWar.htm.

to benefit illicitly from mineral resources. The panel was critical of the connection between the DRC Government and the Zimbabwean Government, including the major mining project in the diamond centre of Mbuji Mayi. Other diamond projects are run as joint ventures involving both DRC and Zimbabwean officials, particularly under the umbrella of the COSLEG stock company. However, the panel argued that projects are run with the support of individuals from the DRC and Zimbabwean military and political elites, which give little or no revenue to the DRC national treasury, while supporting the individual interests of members of those elites. The panel also argued that, because of such arrangements, individuals in the Zimbabwean political and military circles will continue to benefit, even though the formal Zimbabwean military presence from the DRC is in a process of withdrawal. The panel argued that as a result of these transactions, Harare in Zimbabwe became 'a significant illicit diamond-trading centre'.[105]

In the case of Rwanda, it is alleged in the expert panel's report that the regional administration in the Kivu region of the DRC has largely been infiltrated by persons loyal to Rwanda, who will continue to support the efforts of that country to exploit the DRC's mineral wealth, including diamond resources. Profits from resource extraction largely transferred back to the Rwandan Army through its Congo Desk, with only a small amount returning to the coffers of the regional administration.[106]

Similar to the case of Rwanda, the panel argued that the formal withdrawal of the Ugandan military apparatus will not prevent continuing economic gain to individuals who are involved in private businesses in the DRC. In particular, the panel argued that there are a number of rebel groups in Uganda who operate as de facto arms of the Ugandan military, able to represent the interests of the Ugandan elite on an ongoing basis.[107]

In response to the conflict diamonds problem faced by the DRC, the Security Council recommended a number of measures, including the imposition of a diamond embargo on the DRC and Rwanda, Uganda

105 Final Report Panel of Experts on the Illegal Exploitation of Natural Resources and Other Forms of Wealth of the Democratic Republic of the Congo, UN Doc S/2002/1146 (16 October 2002) 6–8.
106 Ibid 14–15.
107 Ibid 19–21.

and Burundi, as well as encouraging the imposition of criminal sanctions against corporations and individuals breaking the embargo. The Security Council also called for an end to trade with certain banks, the seizure of assets of particular individuals, and an arms embargo and suspension of military cooperation with national and rebel military forces operating in the DRC. In contrast to the cases of Sierra Leone, Angola, Côte d'Ivoire, and Liberia, however, sanctions were not immediately imposed on the DRC and its forces of occupation.[108] A 12-month arms embargo on the eastern half of the country was eventually mandated in July 2003, which was extended to the whole of the country in May 2005, along with assets of particular individuals being frozen. Sanctions on commodities such as diamonds were not favoured even as recently as 2007 by a UN expert panel.[109]

In its subsequent 2002 report, the expert panel named not only individuals but a comprehensive list of small enterprises and very large multinational enterprises that were allegedly fuelling the DRC war. The persons and entities named were separated into two categories. The first category, listed in annexes I and II of that report, were targeted for restrictive measures because '[b]y contributing to the revenues of the elite networks, directly or indirectly, those companies and individuals contribute to the ongoing conflict and to human rights abuses'. While noting that those listed would be in violation of the Organisation for Economic Co-operation and Development (OECD) Guidelines for Multinational Enterprises, the individuals and enterprises 'involved in criminal and illicit exploitation' were targeted for particular measures: travel bans on selected individuals identified by the panel; freezing of the personal assets of persons involved in illegal exploitation; barring selected companies and individuals from accessing banking facilities and other financial institutions, and from receiving funding or establishing a partnership or other commercial relations with international financial institutions. The panel

---

108 Report of the Panel of Experts on the Illegal Exploitation of Natural Resources and Other Forms of Wealth of the Democratic Republic of Congo, UN Doc S/2001/357 (12 April 2001) [221]–[225]; Fonseca, A, *Four Million Dead: The Second Congolese War, 1998–2004* (18 April 2004) 68. Available at: www.oocities.org/afonseca/CongoWar.htm.
109 Global Policy Forum, *The Democratic Republic of Congo* (2006). Available at: www.globalpolicy.org/security/issues/kongidx.htm; Interim Report of the Group of Experts on the Democratic Republic of the Congo, Pursuant to Security Council Resolution 1698 (2006), UN Doc S/2007/40 (31 January 2007) [44]–[45].

recommended, however, that a grace period of four to five months be provided for those corporations identified to prove that they had ceased their financial activities in the DRC.[110]

The panel identified a further list of corporations, many large multinationals registered in Europe or the United States, as being presumably less criminally liable in their behaviour, but nevertheless in violation of the OECD Guidelines for Multinational Enterprises. The expert panel reminded national governments of their responsibility to ensure that enterprises in their jurisdiction do not abuse principles of conduct they had adopted as a matter of law, with criminal prosecutions under national legislation being one option for consideration by governments. The panel also recommended that the Security Council establish a monitoring body to oversee action taken by governments in relation to multinational corporations registered in their countries.[111]

Diamond trading corporations featured prominently in both categories identified by the expert panel. In relation to the corporations singled out for direct punitive action, five corporations were exclusively identified with the mining or trading in diamonds. The second list, however, was perhaps more startling for its inclusion of diamond mining giant De Beers and other high-profile corporates such as the mining company Anglo-American, and Barclays Bank. Altogether, 12 exclusively diamond mining/trading enterprises were placed on the category two list.[112]

The UN panel report also called for the universal implementation of the Kimberley Process. Interestingly, they called for the establishment of a permanently staffed secretariat at the international level, and identified the necessity of creating a specialised enforcement organisation within each member country with the authority, knowledge and specialised training necessary to ensure the effectiveness of the Kimberley Process.[113]

---

110 Final Report Panel of Experts on the Illegal Exploitation of Natural Resources and Other Forms of Wealth of the Democratic Republic of the Congo, UN Doc S/2002/1146 (16 October 2002) [176].

111 Ibid [170], [177]–[178].

112 Ibid 5–7, 35–44; Fishman, J L, 'Is Diamond Smuggling Forever?: The Kimberley Process Certification Scheme: The First Step Down the Long Road to Solving the Blood Diamond Trade Problem' (2005) 13 *University of Miami Business Law Review* 217, 217, 222.

113 Final Report Panel of Experts on the Illegal Exploitation of Natural Resources and Other Forms of Wealth of the Democratic Republic of the Congo, UN Doc S/2002/1146 (16 October 2002) 182.

The withdrawal of Ugandan troops in 2002 sparked further problems, particularly in the Ituri area. Despite the deployment of a UN peacekeeping force, the United Nations Mission in the Democratic Republic of Congo (MONUC), the power vacuum created by the withdrawal of troops has now effectively been replaced by elements who defected from the regular forces of the Congolese Army, who now operate under the name of *Mouvement Révolutionnaire du Congo* (MRC). Reports from observers of the situation in 2006 noted that MRC military activities were supported by the illegal exploitation of natural resources in the region.[114] Militia groups operating in the area had a strong incentive to acquire as much gold and diamonds as possible, so as to pre-empt the operation of the Ituri Pacification Committee, whose purpose under the Luanda Agreement is the management of natural resources. As a result, a showdown occurred between Ugandan-backed RCD-ML, *Front des Nationalistes et Intégrationnistes* (FNI), and *Force de Résistance Patriotique en Ituri* (FRPI) on the one hand, and the Rwandan-backed *Union des Patriotes Congolais* (UPC), supported by the Congolese Liberation Movement (MLC) and the RCD-National. The Ituri conflict was further complicated by an ethnic dimension, notably that the Ugandan-backed militias were composed primarily of people of Lendu ethnic background, while the UPC was primarily composed of persons of Hema ethnicity. The ethnic dimension sparked fears of a possible genocide, which fortunately didn't eventuate.[115]

Human rights violations in the Ituri region, however, attracted the attention of the international criminal justice system, leading to indictments and the first arrests by the newly established ICC.[116] Thomas Lubanga Dyilo, leader of the UPC, was arrested on the same day as the issuance of the indictment against him on 17 March 2006, with the assistance of Congolese authorities, French armed forces, and MONUC forces.[117] The indictment against Lubanga alleges that the UPC, under the leadership of Lubanga, seized control of Bunia and parts of

---

114 Africa Initiative Programme and Forum on Early Warning and Early Response (FEWER-Africa), 'Elections and Security in Ituri: Stumbling Blocks and Opportunities for Peace in the Democratic Republic of Congo' (Report, 13 June 2006) 4–27.

115 Fonseca, A, *Four Million Dead: The Second Congolese War, 1998–2004* (18 April 2004) 81–82. Available at: www.oocities.org/afonseca/CongoWar.htm.

116 Ibid 82.

117 International Criminal Court, Office of the Prosecutor, 'Issuance of a Warrant of Arrest against Thomas Lubanga' (Press Release, ICC-OTP-20060302-126-En, 2 March 2006); International Criminal Court, Office of the Prosecutor, 'First Arrest for the International Criminal Court' (Press Release, ICC-CPI-20060302-125-En, 17 March 2006).

Ituri in Orientale Province in 2002, resulting in the death of more than 8,000 civilians and the displacement of 600,000 others. The indictment also alleges that the UPC took young children from their families and forced them to join the UPC military forces, which constitutes a crime under Articles 8(2)(b)(xxvi) and 8(2)(e)(vii) *Rome Statute*.[118] The ICC followed up its first arrest with a second on 18 October 2007, also in relation to alleged crimes in the Ituri area. The prosecutor alleged that Germain Katanga was a senior commander of the FRPI and was responsible for the massacre of hundreds of civilians in the village of Bogoro on 24 February 2004, as well as crimes of sexual violence and enlisting child soldiers.[119] The third arrest, also relating to the Ituri situation, occurred on 7 February 2008. Mathieu Ngudjolo Chui had allegedly been the leader of the FNI at the time that the militia, along with the FRPI, attacked the Bogoro village and therefore faced largely parallel charges to Katanga.[120]

The invasion of the DRC by its neighbours has also resulted in a number of cases before the International Court of Justice, which adjudicates international legal disputes between nations. In Armed Activities on the Territory of the Congo, the DRC claimed that the invasion of Congolese territory by Burundian, Ugandan, and Rwandan troops on 2 August 1998 constituted a violation of its sovereignty and its territorial integrity, as well as a threat to peace and security in Central Africa. The issue of illegal seizure of assets was also raised by the DRC. The International Court of Justice found that it had jurisdiction to consider issuing preliminary measures against Uganda.[121] A further case before the International Court followed from the issuance of an

---

118 International Criminal Court, Office of the Prosecutor, 'Issuance of a Warrant of Arrest against Thomas Lubanga' (Press Release, ICC-OTP-20060302-126-En, 17 March 2006).

119 International Criminal Court, 'Case Information Sheet for The Prosecutor v Germain Katanga, ICC-01/04-01/07'. Available at: www.icc-cpi.int/en_menus/icc/situations%20and%20 cases/situations/situation%20icc%200104/related%20cases/icc%200104%200107/Pages/ democratic%20republic%20of%20the%20congo.aspx.

120 International Criminal Court, 'Case Information Sheet for The Prosecutor v Mathieu Ngudjolo Chui, ICC-01/04-02/12', www.icc-cpi.int/en_menus/icc/situations%20and%20cases/ situations/situation%20icc%200104/related%20cases/ICC-01-04-02-12/Pages/default.aspx. Human rights violations and ongoing conflict have also occurred in the Rwandan-influenced regions of North and South Kivu, although a recent peace deal with General Nkunda, leader of a RCD-Goma splinter group, may herald the way to greater stability. See British Broadcasting Corporation, 'Eastern Congo Peace Deal Signed' (23 January 2008). Available at: www. globalpolicy.org/security/issues/congo/2008/0123gomadeal1.htm.

121 Gray, C, 'The Use and Abuse of the International Court of Justice: Cases Concerning the Use of Force After Nicaragua' (2003) 14 *European Journal of International Law* 867, 878–880.

arrest warrant by Belgium on 11 April 2000 against the incumbent DRC Foreign Minister, Mr Abdulaye Yerodia Ndambasi, alleging his commission of war crimes and crimes against humanity, pursuant to a Belgian criminal statute. The International Court ruled that the customary international rule providing immunity to incumbent foreign ministers against civil and criminal proceedings rendered the Belgian arrest warrant unlawful under international law. The court noted that this rule would not extend to proceedings initiated against former foreign ministers or proceedings before international criminal courts.[122]

Recent statistics indicate that the legitimate diamonds trade in the DRC is recovering, indicating some degree of success of the Kimberley Process. In 2005, official exports were valued at $642 million, constituting a 62.5 per cent increase on the previous year. New and independent valuation was partly responsible for the increase, as well as effective implementation of the Kimberley Process and the expulsion of Congo-Brazzaville from the process in 2004. The expulsion of Congo-Brazzaville, a neighbour of the DRC, resulted in an increase in the legal trade because it was recognised that Congo-Brazzaville was being used as a back door to avoid the Kimberley Process Certification System. As a result of its expulsion from the system, Kimberley Process members were prohibited from purchasing that country's diamonds, meaning that much of the diamond trade reverted to legitimate channels in the DRC. It should be noted, however, that there was still an estimated US$350 million leaving the country illicitly, indicating the scale of work still to be done with the industry.[123] Given the example of conflicts in other nations, such periods of relative peace carry with them the threat of re-armament through the exploitation of conflict diamonds and other natural resources.[124]

---

122 Golden Gate University School of Law, 'Annex: Arrest Warrant of 11 April 2000 (Democratic Republic of the Congo v. Belgium): International Court of Justice 14 February 2002' (2002) 8 *Annual Survey of International and Comparative Law* 151, 159–161.

123 Global Witness and Partnership Africa Canada, 'Rich Man, Poor Man, Development Diamonds and Poverty Diamonds: The Potential for Change in the Artisanal Alluvial Diamond Fields of Africa' (Report, October 2004) 25–32. The Republic of Congo was re-admitted in 2007 to the Kimberley Process. Hennessy, S, *Congo's Diamond Industry Let Back into Kimberley Process* (9 November 2007) Global Policy Forum. Available at: www.globalpolicy.org/security/issues/congo/2007/1109drckp.htm.

124 Malamut, S A, 'A Band-Aid on a Machete Wound: The Failures of the Kimberley Process and Diamond-Caused Bloodshed in the Democratic Republic of Congo' (2005) 29 *Suffolk Transnational Law Review* 25, 43.

In the period since 2005, although much of the DRC has enjoyed relative peace, conflict fuelled by natural resources is an ongoing feature of the disturbed north-eastern provinces. Even though minerals other than diamonds are the focus of reports by UNSC expert panels in 2007 and 2008, it is known that diamond deposits also exist in these conflict areas, so it may be premature to argue that conflict diamonds are no longer a feature of the DRC conflicts. A further development occurred with the release in August 2010 of the DRC 'Mapping Exercise' by the UN High Commissioner for Human Rights. The exercise sought to provide preliminary evidence of international crimes committed from 1993 to 2003 in the context of the Congolese war, and the report features a section exploring the connection between resources and conflict, and the role it played during the war. It is possible that a UN international tribunal may be established to prosecute persons most responsible for committing war crimes, crimes against humanity, and genocide during this period.[125]

Côte d'Ivoire

The conflict diamonds problem arose dramatically with the outbreak of civil war in Côte d'Ivoire, also known as the Ivory Coast, in September 2002.[126] According to a UNSC investigation, the Forces Nouvelle militia in the north of that country, who went to war following discontent with a controversial election process, have been largely funded through the illegal diamond trade. The resurgence of the problem, in the era where the Kimberley Process Certification Scheme is active, has proven a significant test of the resolve of the international community in its efforts to combat the problem.[127]

---

125 Interim Report of the Group of Experts on the Democratic Republic of the Congo, Pursuant to Security Council Resolution 1698 (2006), UN Doc S/2007/40 (31 January 2007); Final Report of the Group of Experts on the Democratic Republic of the Congo, UN Doc S/2008/773 (21 November 2008); United Nations Office of the High Commissioner for Human Rights, Report of the Mapping Exercise documenting the Most Serious Violations of Human Rights and International Humanitarian Law Committed within the Territory of the Democratic Republic of the Congo between March 1993 and June 2003 (August 2010) 349–367.

126 British Broadcasting Corporation, 'Q&A: Ivory Coast's Crisis', *BBC News: World Edition* (17 January 2006). Available at: news.bbc.co.uk/2/hi/africa/3567349.stm.

127 Kofele-Kale, N, 'The Global Community's Role in Promoting the Right to Democratic Governance and Free Choice in the Third World' (2005) 11 *Law and Business Review of the Americas* 205, 235; Wexler, Pamela, *The Kimberley Process Certification Scheme on the Occasion of its Third Anniversary: An Independent Commissioned Review* (Review Report submitted to the Ad Hoc Working Group on the Review of the Kimberley Process, February 2006) 7.

Côte d'Ivoire had experienced widespread stability and prosperity for more than three decades after independence under the leadership of its first president, Félix Houphouët-Boigny. It was considered to have the fourth-largest economy in sub-Saharan Africa. However, military rule from 1999 to 2000 under Robert Guéï, general political unrest under President Laurent Gbagbo, and the outbreak of civil conflict in 2002 led to widespread atrocities, allegedly conducted by both government and rebel forces, including political killings, massacres, disappearances, and numerous incidents of torture. While a peace agreement was brokered between the two sides in 2003, widespread impunity and a political and social climate fuelled by intolerance and xenophobia have caused fears that the hostilities will resume. UN peacekeepers have patrolled a buffer zone separating the rebel-held north and the government-controlled south, but political efforts to reunite the nation have not been successful to date. In September 2003, Côte d'Ivoire requested that the ICC accept jurisdiction over crimes committed on its territory since 19 September 2002.[128]

Like the situation that affected Sierra Leone previously, the establishment by rebel militia of a war economy based on diamonds served to entrench the power of the militia and to delay the satisfactory resolution of the conflict, which centres around ethnic grievances and issues of political representation.[129] A UNSC expert panel reported that high levels of funds were reaching the northern-based militia through diamond sales, and that strict measures were required for this to be controlled. The report of the expert group resulted in a Security Council ban on imports of rough diamonds from Côte d'Ivoire, pursuant to Resolution 1643 (2005).[130] The expert group had conducted its activities through the support of the current peacekeeping mission

---

128 Punyasena, W, 'Conflict Prevention and the International Criminal Court: Deterrence in a Changing World' (2006) 14 *Michigan State Journal of International Law* 39, 64–65; Petrova, P, 'The Implementation and Effectiveness of the Kimberley Process Certification Scheme in the United States' (2006) 40 *International Lawyer* 945, 945.

129 British Broadcasting Corporation, 'Q&A: Ivory Coast's Crisis', *BBC News: World Edition* (17 January 2006). Available at: news.bbc.co.uk/2/hi/africa/3567349.stm.

130 SC Res 1643, UN SCOR, 5327th mtg, S/RES/1643 (15 December 2005) (Côte d'Ivoire).

in Côte d'Ivoire, United Nations Operation in Côte d'Ivoire (UNOCI).[131] The responses of the UNSC and the Kimberley Process to this issue is discussed in more detail in chapters 4 and 5.

Côte d'Ivoire is a minor producer of rough diamonds and has been a participant in the Kimberley Process Certification Scheme since its inception. In a letter dated 13 October 2004, the Minister of Mines and Energy of Côte d'Ivoire informed the Chair of the Kimberley Process that all exports of rough diamonds from Côte d'Ivoire were prohibited on the basis of a ministerial order issued on 19 November 2002. As there are no significant diamond deposits in the government-controlled south, this measure seeks to cut off the northern rebels from international diamond markets.[132]

The group of experts investigated the production and export of rough diamonds from Seguela, Bobi, and Diarabla localities in northern Côte d'Ivoire in July 2005 and obtained credible information about diamond production in the region. The group visited the area and received information that significant artisanal production of rough diamonds was occurring along a number of small streams between the villages, but that semi-industrial or industrial mining techniques were not used.[133] Based on the number of active pits and workers, the expert group estimated that production was at a level of 300,000 carats per year, equal to production prior to the conflict. The revenue accrued by the groups controlling the production is many millions of US dollars per year.[134] Rather than mining diamonds themselves, Forces Nouvelles were raising their revenue through systems of taxation. The militia appropriated the local diamond bureaucracy, and so were directly receiving taxes on diamond production that had previously gone to the national government. The militia also employed

131 Report of the Group of Experts Submitted Pursuant to Paragraph 7 of Security Council Resolution 1584 (2005) Concerning Côte d'Ivoire, UN Doc S/2005/699 (7 November 2005); Update Report of the Group of Experts Submitted Pursuant to Paragraph 2 of Security Council Resolution 1632 (2005) Concerning Côte d'Ivoire, UN Doc S/2006/204 (31 March 2006).
132 Report of the Group of Experts Submitted Pursuant to Paragraph 7 of Security Council Resolution 1584 (2005) Concerning Côte d'Ivoire, UN Doc S/2005/699 (7 November 2005) 17–20.
133 Ibid.
134 Ibid.

an indirect approach, through imposing a tax on those who purchased automobiles and motorcycles with revenue from diamond mining, or alternatively on persons seeking to access diamond mining areas.[135]

Like the situation in other affected countries, conflict diamonds have been smuggled out of Côte d'Ivoire through neighbouring states. In this case, the countries in question are apparently Guinea — a Kimberley Process participant — and Mali — a Kimberley Process applicant at the relevant time — after which the diamonds have been distributed to international markets, such as Antwerp, Dubai, or Tel Aviv. It is interesting to note that, unlike the situation of Republic of Congo, neither Mali nor Guinea was expelled from the Kimberley Process in response to their alleged role.[136]

According to a UN expert's report in 2010, diamonds from the Seguela and Tortiya regions of northern Côte d'Ivoire continue to be smuggled through neighbouring Burkina Faso, Guinea, Liberia, and Mali, which are unable or unwilling to enforce the diamond embargo. It is unclear, however, if the trade is connected to the most recent wave of violence in the country, in which outgoing President Laurent Gbagbo chose violent struggle over peaceful transition to the administration of newly elected President Alassane Ouattara. Gbagbo was ultimately arrested for international crimes related to the brief armed struggle occurring from latter part of 2010 to early 2011.[137] His arrest marked the end of the brief civil war.[138] His case at the ICC was joined to that of Ble Goude on 11 March 2015, and the trial commenced on 28 January 2016.[139]

---

135 Ibid; Update Report of the Group of Experts Submitted Pursuant to Paragraph 2 of Security Council Resolution 1632 (2005) Concerning Côte d'Ivoire, UN Doc S/2006/204 (31 March 2006) 10.
136 Report of the Group of Experts Submitted Pursuant to Paragraph 7 of Security Council Resolution 1584 (2005) Concerning Côte d'Ivoire, UN Doc S/2005/699 (7 November 2005) 17–20.
137 Report of the Group of Experts on Côte d'Ivoire pursuant to Paragraph 11 of Security Council Resolution 1946 (2010), UN Doc S/2011/272 (17 March 2011) 55–61; 'Diamonds: Côte d'Ivore' (February 2011) 41(12) Africa Research Bulletin, 'Diamonds: Zimbabwe' (2011) 41(12) February *Africa Research Bulletin: Economic, Financial and Technical Series* 18960; Dawn News, 'Blood Diamond Fears in Ivory Coast Political Duel' (28 December 2010). Available at: www.dawn.com/2010/12/28/blood-diamond-fears-in-ivory-coast-political-duel.html; McClanahan, P, *As Ivory Coast's Gbagbo Holds Firm, 'Blood Diamonds' Flow for Export* (23 January 2011) ReliefWeb. Available at: reliefweb.int/node/381665.
138 British Broadcasting Corporation, 'Gbagbo Held After Assault on Residence' (11 April 2011). Available at: www.bbc.com/news/world-africa-13039825.
139 The International Criminal Court, *The Prosecutor vs. Laurent Gbagbo and Ble Goude*, ICC-02/11-01/15. Available at: www.icc-cpi.int/cdi/gbagbo-goude/Pages/default.aspx.

Zimbabwe

Alluvial diamonds were discovered in the Chiadzwa district of Marange, eastern Zimbabwe, in June 2006. The diamond fields stretch over 66,000 hectares and, although estimates of the reserves contained in this area vary wildly, some have gone so far as to suggest that it could be home to one of the world's richest diamond deposits. Illicit production from the region, up to the end of 2008, was valued at approximately US$150 million. Over the past four years, Marange has been plagued by horrific human rights abuses by state security agencies against diamond diggers and local communities, resulting in hundreds of deaths, and many more cases of assault, rape, arbitrary detention, and forced labour. [140]

From early 2007, police officers stationed in the fields began forcing miners to work in syndicates under their control, demanding bribes, and beating or killing anyone else they found mining in the area. The violence reached a peak in October 2008 with the arrival of the army, and the launch of Operation Hakudzokwi or 'You will not return'. This operation appeared to have two goals: to ensure control of the diamond deposits for the Zanu Patriotic Front elite, and to reward the army for its loyalty to this clique. More than 800 soldiers were deployed alongside helicopter gunships, killing over 200 people.[141]

Following this operation, soldiers took over mining syndicates previously run by the police, and forced local people, including children, to mine for them. The military was also central in facilitating smuggling diamonds out of Zimbabwe to neighbouring countries,

---

140 Human Rights Watch, 'Diamonds in the Rough: Human Rights Abuses in the Marange Diamond Fields of Zimbabwe' (Report, June 2009) 3–4; Partnership Africa Canada, 'Zimbabwe, Diamonds and the Wrong Side of History' (Report, March 2009) 3, 5; Partnership Africa Canada, 'Diamonds and Clubs: The Militarised Control of Diamonds and Power in Zimbabwe' (Report, June 2010) 18; Global Witness, 'Return of the Blood Diamond: The Deadly Race to Control Zimbabwe's New-found Diamond Wealth' (Report, 14 June 2010) 6–8.
141 Human Rights Watch, 'Diamonds in the Rough: Human Rights Abuses in the Marange Diamond Fields of Zimbabwe' (Report, June 2009) 3–4; Partnership Africa Canada, 'Zimbabwe, Diamonds and the Wrong Side of History' (Report, March 2009) 7–8; Partnership Africa Canada, 'Diamonds and Clubs: The Militarised Control of Diamonds and Power in Zimbabwe' (Report, June 2010) 18; Global Witness, 'Return of the Blood Diamond: The Deadly Race to Control Zimbabwe's New-found Diamond Wealth' (Report, 14 June 2010) 6–8. There was an initial KP review visit in May to June 2007, prior to the worst violence. The review focused on the country's regulatory framework and noted that the contentious fields were under government control. The KP review was criticised for failing to highlight ongoing human rights abuses. Partnership Africa Canada, 'Diamonds and Human Security: Annual Review 2008' (Annual Report, October 2008) 23–24.

including Mozambique and South Africa. Once again, civilians found digging for diamonds independently of the syndicates were severely beaten or killed as a warning to others.[142]

A Kimberley Process Review Mission was eventually sent to the country in June 2009 to investigate the violence and assess compliance with KP standards. The mission found evidence of grave human rights abuses, armed soldiers managing syndicates of miners, and a 'smuggling operation that enables rough diamonds to flow from Zimbabwe outside the KPCS ... largely operated and maintained by official entities'. This finding alone — that state agents were running diamond smuggling operations to Mozambique, a non-KP participant — should be grounds for expulsion from the scheme.[143]

The review team, made up of government, NGO, and industry representatives, 'identified several areas in which Zimbabwe [is] non-compliant with the minimum requirements of the KPCS', and recommended that the country be suspended from the scheme for at least six months.[144]

In a press conference held at the end of the visit, the mission's leader, Liberian Deputy Minister of Mines Kpandel Fayia, made an impassioned plea to the Zimbabwean authorities:

> Minister, on the issue of violence against civilians, I need to be clear about this. Our team was able to interview and document the stories of tens of victims, observe their wounds, scars from dog bites and batons, tears, and on-going psychological trauma. I am from Liberia, Sir; I was in Liberia throughout the 15 years of civil war; and I have experienced too much senseless violence in my lifetime, especially connected to diamonds. In speaking with some of these people, Minister, I had to leave the room. This has to be acknowledged and it has to stop.[145]

---

142 Human Rights Watch, 'Diamonds in the Rough: Human Rights Abuses in the Marange Diamond Fields of Zimbabwe' (Report, June 2009) 3–4; Partnership Africa Canada, 'Zimbabwe, Diamonds and the Wrong Side of History' (Report, March 2009) 7–8; Partnership Africa Canada, 'Diamonds and Clubs: The Militarised Control of Diamonds and Power in Zimbabwe' (Report, June 2010) 15–16; Global Witness, 'Return of the Blood Diamond: The Deadly Race to Control Zimbabwe's New-found Diamond Wealth' (Report, 14 June 2010) 6–8.
143 Partnership Africa Canada, 'Diamonds and Human Security: Annual Review 2009' (Annual Report, 2009) 19; Global Witness, 'Return of the Blood Diamond: The Deadly Race to Control Zimbabwe's New-found Diamond Wealth' (Report, 14 June 2010) 6–8.
144 Global Witness, 'Return of the Blood Diamond: The Deadly Race to Control Zimbabwe's New-found Diamond Wealth' (Report, 14 June 2010) 6–8.
145 Ibid.

The situation in the Marange diamond fields remains critical. The Zimbabwean authorities claim that the joint venture companies they have recently established and given permits to mine in Marange will help regulate the diamond sector and improve standards. However, these companies are only operating in around 3 per cent of the diamond fields, with the remaining 97 per cent under the control of the army.[146]

The widespread smuggling of Marange diamonds out of Zimbabwe persists, and the army continues to operate syndicates of miners as a means of capturing the proceeds of this illegal trade.[147]

There have been some cases of enforcement action by national governments in relation to smuggled Zimbabwean diamonds. On 20 September 2008, India's Directorate of Revenue Intelligence (DRI) apprehended two Lebanese nationals, named as Yusuf Oselli and Robar Hussain, from a hotel in Surat, the centre of India's diamond industry. They found rough diamonds weighing 3,600 carats and valued at almost $800,000. The pair said they had brought the diamonds from Zimbabwe, and that they had made several earlier runs. The men, who did not have a Kimberley Process certificate or any other documentation for the diamonds, told the DRI that they had carried the diamonds through Dubai, landing undetected at Mumbai on 15 September.[148]

In October 2008, Dubai Customs discovered bags of diamonds wrapped around the body of a Zimbabwean woman transiting in Dubai. The diamonds weighed 53,500 carats and were valued at US$1.2 million. The information was carried in local and international media, including a photograph that shows that the diamonds resemble

---

146 Ibid.
147 Simpson, John, 'Profiting from Zimbabwe's "blood diamonds"', *BBC News* (online), 20 April 2009. Available at: news.bbc.co.uk/2/hi/africa/8007406.stm; Partnership Africa Canada, 'Zimbabwe, Diamonds and the Wrong Side of History' (Report, March 2009) 7–8; Global Witness, 'Return of the Blood Diamond: The Deadly Race to Control Zimbabwe's New-found Diamond Wealth' (Report, 14 June 2010) 6–8. Mugabe's wife has personally benefited from the mining according to some reports. Herskovitz, Jon, 'Zimbabwe Govt, c.bank in Blood Diamond Trade: WikiLeaks', *Reuters: Africa* (9 December 2010). Available at: af.reuters.com/article/topNews/idAFJOE6B80F820101209.
148 Partnership Africa Canada, 'Zimbabwe, Diamonds and the Wrong Side of History' (Report, March 2009) 12.

those originating from Marange. There is no information on whether the woman was charged, her travel origin or destination, or the disposition of the diamonds.[149]

Although access to the Marange diamond fields has been severely restricted, testimony gathered from victims by local civil society representatives shows that serious human rights abuses, including assault and rape, are still being committed by the army and the police.[150]

In March 2010, the Centre for Research and Development, an NGO based in Mutare, the provincial capital, identified 26 victims of abuse in the diamond fields and the surrounding area, including two cases of rape, and one of a woman being beaten so severely she was left partially blind. In April, the same NGO recorded 24 cases of assault by the security forces against civilians.[151]

Some local experts believe that the actual number of assaults is much higher, but that people are too afraid to report abuses for fear of further harassment. The researchers also note that the violence often precedes visits to the area by important government delegations — an apparent attempt to clear the area of miners before the visitors arrive.[152]

Despite the continued violence, Zimbabwe remains a member of the Kimberley Process, the international certification body set up to prevent diamond-fuelled violence and abuses. The failure of KP member states to agree to suspend Zimbabwe has prompted deep concern among some KP participants and observers, who have begun to question the future of the scheme.[153]

In 2009, a compromise approach to the Zimbabwe issue was agreed to by the KP, whereby the KP sent an official monitor to the Chiadzwa region. Zimbabwe was given six months to fall in line with international trade standards, pursuant to a 'Joint Work Plan' that included the demilitarisation of the diamond fields. However, this has not happened, and there have been ongoing reports of smuggling and harassment by military officers. Despite this, the KP allowed two auctions of stockpiled

149 Ibid 12–13.
150 Global Witness, 'Return of the Blood Diamond: The Deadly Race to Control Zimbabwe's New-found Diamond Wealth' (Report, 14 June 2010) 6–8.
151 Ibid.
152 Ibid.
153 Ibid.

diamonds in 2010. The sales were meant to pave the way for full exports to resume, but KP members have not reached the necessary consensus on whether to allow full exports to resume. The KP has contradicted claims by the Zimbabwean Mines Minister in January 2011 that it has been given permission to make further official sales.[154]

Central African Republic

The Central African Republic (CAR) conflict is an ongoing civil war in the CAR between the *Séléka* rebel coalition and government forces, which began on 10 December 2012. The conflict arose after rebels accused the government of President François Bozizé of failing to abide by peace agreements signed in 2007 and 2011. Many of the rebel groups had been involved in the CAR Bush War.

Rebel forces known as *Séléka* (meaning 'union' in the Sango language) captured many major towns in the central and eastern regions of the country at the end of 2012. *Séléka* comprises two major groups based in north-eastern CAR: the Union of Democratic Forces for Unity (UFDR) and the Convention of Patriots for Justice and Peace (CPJP), but also includes the lesser known Patriotic Convention for Saving the Country (CPSK). Two other groups based in northern CAR — the Democratic Front of the Central African People (FDPC) and the Chadian group Popular Front for Recovery (FPR) — also announced their allegiance to the *Séléka* coalition.

Chad, Gabon, Cameroon, Angola, South Africa, the DRC and the Republic of Congo sent troops as part of the Economic Community of Central African States' FOMAC force to help the Bozizé government hold back a potential rebel advance on the capital, Bangui. However, the capital was seized by the rebels on 24 March 2013 at which time President Bozizé fled the country, and the rebel leader Michel Djotodia

---

154 Africa Research Bulletin, 'Diamonds: Zimbabwe' (2011) 41(12) February *Africa Research Bulletin: Economic, Financial and Technical Series* 18960, 18960–18961; Global Witness, *Industry Must Refuse Zimbabwe Diamonds Certified by Rogue Monitor* (16 November 2010). Available at: www.globalwitness.org/library/industry-must-refuse-zimbabwe-diamonds-certified-rogue-monitor; Global Witness, *Conflict Diamond Scheme Must Resolve Zimbabwe Impasse* (5 November 2010). Available at: www.globalwitness.org/library/conflict-diamond-scheme-must-resolve-zimbabwe-impasse.

declared himself president. Due to the presence of diamonds in the country, the UNSC imposed a diamond embargo, which was backed by action from the Kimberley Process.

India

Beyond the current mandate of the Kimberley Process, which focuses on the connection of the diamond trade to human rights violations reaching the level of international crimes, lies the potential for addressing a range of human rights issues that, although serious, do not constitute international crimes. One of these issues is the use of child labour, which is a major problem not only in the artisanal mining sector, but also in the cutting and polishing industry, which is dominated by India.

The skills required for cutting and polishing diamonds are passed down by workers from generation to generation, or are picked up in the traditional master–apprentice relationship. Of the four Cs — colour, clarity, carat, and cut — nature dictates the first three aspects; the cut, often considered the benchmark against which a diamond's beauty is judged, is the only factor determined by the human hand. It is a practice requiring great expertise. However, behind the glittering world of India's diamond-cutting industry lie practices of exploitation and child labour. India enjoys a near monopoly in the diamond-cutting industry, but low wages and the easy availability of labour is what keeps the industry profitable. India gets a lot of small diamonds to cut and polish. The detailed nature of the work and the repetitive strain of cutting and polishing these tiny specks of stones make it labour-intensive and often unhealthy. There is a lot of dust from the ground diamonds that doesn't always get filtered out of the crowded factory rooms, and proves harmful for workers' health. These small stones often need sharp eyes and deft hands, and children are often highly prized in the trade, able to cut even 'half-pointer' diamonds, noting that 100 points make a carat, which is one-fifth of a gram.[155]

---

155 Hussain, S S, 'A Diamond's Journey: Grime Behind the Glitter?', *World News* (26 June 2009). Available at: www.msnbc.msn.com/id/15842528; Brilliant Earth, *Conflict Diamond Issues* (2007). Available at: www.brilliantearth.com/conflict-diamond-child-labor/.

Child labour is illegal in India, but remains widespread. By conservative estimates, 13 million children work in India, many in hazardous industries. According to one estimate, up to 100,000 children, in the age group 6–14 years, work in the diamond industry in Surat, cutting and polishing diamond chips.[156]

Children are engaged as apprentices, with learning the trade taking from five to seven years. During the first two years of an apprenticeship children receive little or no remuneration, working for 10 hours a day. After two years, a child worker is paid about $1.70 per month. Studies by noted academic Neeta Burra revealed that more than 30 per cent of these children get tuberculosis due to unhygienic conditions, overcrowding, and malnutrition. Major health issues include body aches, and finger tips grazed by the polishing disc. Child labour continues to be a major problem, despite efforts to stop the practice and generally improve working conditions amongst some in the cutting industry.[157]

Concluding Remarks

This book considers two main research questions:

1. Has the conflict diamonds governance system achieved its objectives?
2. Does consideration of the system from the perspective of the networked pyramid model provide descriptive or normative insights?

Before a fitting response to these questions can be formulated, however, it is necessary to examine carefully the nature of the conflict diamonds problem. The conflict diamonds trade stands in contrast to the legitimate diamond trade, with rough diamond production in the legal industry during 2006 valued at US$11.5 billion, and exports valued at $US37.7 billion. Over the past century, large-scale

---

156 Hussain, S S, 'A Diamond's Journey: Grime Behind the Glitter?', *World News* (26 June 2009). Available at: www.msnbc.msn.com/id/15842528; Brilliant Earth, *Conflict Diamond Issues* (2007). Available at: www.brilliantearth.com/conflict-diamond-child-labor/.
157 Hussain, S S, 'A Diamond's Journey: Grime Behind the Glitter?', *World News* (26 June 2009). Available at: www.msnbc.msn.com/id/15842528; Brilliant Earth, *Conflict Diamond Issues* (2007). Available at: www.brilliantearth.com/conflict-diamond-child-labor/.

diamond production in the legitimate trade has been characterised by the monopolistic behaviour of the giant De Beers multinational corporation, and a high level of secrecy. At the same time, the low-tech mining of alluvial rough diamond deposits has provided insurgent groups with a commodity that is easy to smuggle, given the small size and very high value of such diamonds, which are considered the most concentrated form of wealth in the world. These characteristics have facilitated the emergence of conflict diamonds used to finance grave human rights violations and armed conflict against established governments. The issue was first highlighted by NGOs in the context of the civil war in Angola, where rebel group UNITA was able to fund its military campaign from diamond sales between 1992 and 1998. The diamond trade prolonged the civil war, which resulted in the loss of around 1 million lives, and reportedly involved such war crimes as the indiscriminate shelling of civilians. Around the same time, civil war fuelled by conflict diamonds also emerged in Sierra Leone. The infamous terror tactics of the RUF, which included the amputation of limbs and the conscription of child soldiers, resulted in a renewed focus on the connection between conflict diamonds and egregious human rights violations. More recently, conflict diamonds have fuelled ongoing civil war situations and human rights violations in the DRC and Côte d'Ivoire. The heavy-handed response of the Zimbabwean Government to the management of artisanal diamond mines in the Marange region represents the latest manifestation of the conflict diamonds problem.

3

The Kimberley Process: Did the Lion Roar?

In a world of failures, this is a story about NGO campaigning, corporate social responsibility and diplomacy that still has a chance of working, not just to end and prevent conflict, but to turn diamonds with secrets and blood in their pedigree into an engine of development and hope in places where these virtues are in tragically short supply.

Ian Smillie, conflict diamonds expert, referring to the KP[1]

Chapter Overview

This chapter discusses the Kimberley Process Certification Scheme (KP), which is the centrepiece of the conflict diamonds governance system. The chapter gives an overview of the Kimberley Process, and discusses its operation, including the way it accepts new members, its annual reporting, and its distinctive peer-review mechanism. In discussing peer review, there is a particular focus on the KP's management of cases of serious non-compliance. The role of industry in the KP is discussed, including self-regulation, as is the role of non-governmental organisations (NGOs), with a particular focus on

---

1   Smillie, I, *Blood on the Stone: Greed, Corruption and War in the Global Diamond Trade* (Anthem Press, 2010) 207.

the continuing role of NGOs beyond the KP, including the Diamond Development Initiative International (DDII). The implementation of KP obligations at the national level is the final focus of analysis, including consideration of regulatory options under national legislation in Angola, the Democratic Republic of Congo (DRC), Côte d'Ivoire, Sierra Leone, Australia, and the Netherlands.

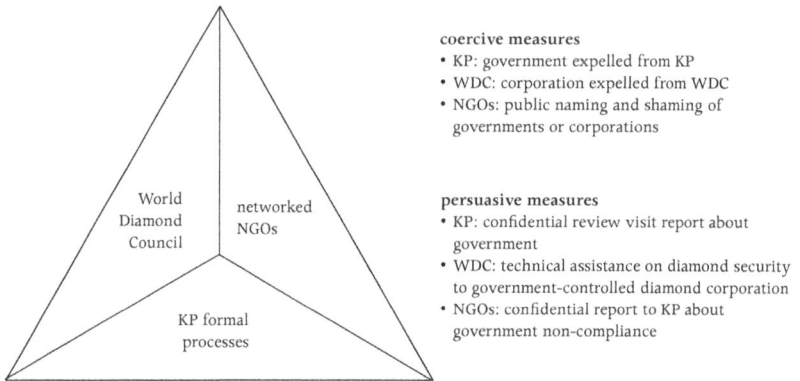

coercive measures
• KP: government expelled from KP
• WDC: corporation expelled from WDC
• NGOs: public naming and shaming of governments or corporations

persuasive measures
• KP: confidential review visit report about government
• WDC: technical assistance on diamond security to government-controlled diamond corporation
• NGOs: confidential report to KP about government non-compliance

Figure 3.1: Kimberley Process: Networked Regulators Operating in Parallel
Source: Author's research.

Overview of the Kimberley Process

The Kimberley Process Certification Scheme is the primary response by the international community to overcome the conflict diamonds problem. As discussed in Chapter 2, conflict diamonds are rough diamonds that are associated with conflict and gross human rights violations. The Kimberley Process aims to distinguish the legitimate rough diamond trade from the illegal trade through the use of a system of certification at the point of export. The central obligation of governmental participants in the KP is to ensure that all conflict-free rough diamond goods are certified at the point of export, and to ensure that only legally certified rough diamonds are imported into the country.[2]

The Kimberley Process was launched in May 2000 in the city of Kimberley, South Africa. It began as a consultative process involving governments, business, and NGOs, later becoming a negotiating process that culminated in the adoption of the Kimberley Process core document at a ministerial meeting in Interlaken, Switzerland, in November 2002.[3] Participating nations agreed to implement the provisions of that agreement by 1 January 2003. The agreement took just 30 months to negotiate, which has been described as blazing speed by UN standards.[4] The Kimberley Process received the formal endorsement of the UN General Assembly (UNGA)

---

2   Kimberley Process, *The Kimberley Process Certification Scheme* (Core Document, 2002) ss II, III. See also Wright, C, 'Tackling Conflict Diamonds: The Kimberley Process Certification Scheme' (2004) 11(4) *International Peacekeeping* 697, 699–700; Fishman, J L, 'Is Diamond Smuggling Forever?: The Kimberley Process Certification Scheme: The First Step Down the Long Road to Solving the Blood Diamond Trade Problem' (2005) 13 *University of Miami Business Law Review* 217, 218–219. Although the central objective of the KP is to exclude trading in rough diamonds from non-members, the 2008 plenary discussed the possibility of a mechanism to provide for trade between participants and certain authorised non-participants (in particular, Turkey). Kimberley Process Chair, 'Report on the Intersessional Meeting of the Working Groups of the Kimberley Process Certification Scheme' (New Delhi, India, 19 June 2008).

3   Kimberley Process, *The Kimberley Process Certification Scheme* (Core Document, 2002); Kimberley Process, *Interlaken Declaration of 5 November 2002 on the Kimberley Process Certification Scheme for Rough Diamonds* (5 November 2002). Preliminary meetings were held prior to finalising the core document, in Namibia: Kimberley Process, 'Kimberley Process Meeting and Technical Workshop' (Report on Preliminary Meeting, Windhoek, Namibia, 13–16 February 2001); Belgium: Kimberley Process, 'Kimberley Process Meeting: Final Communique' (Report on Preliminary Meeting, Brussels, Belgium, 25–27 April 2001); the Russian Federation: Kimberley Process, 'Kimberley Process: Final Communique' (Report on Preliminary Meeting, Moscow, Russia, 3–4 July 2001); England: Kimberley Process, 'Kimberley Process: Meeting in Twickenham' (Report on Preliminary Meeting, Twickenham, England, 11–13 September 2001); and Canada: Kimberley Process, 'Kimberley Process Meeting Final Communique' (Report on Preliminary Meeting, Ottawa, Canada, 18–20 March 2002). See also Wallis, A, 'Data Mining: Lessons from the Kimberley Process for the United Nations' Development of Human Rights Norms for Transnational Corporations' (2005) 4(2) *Northwestern Journal of International Human Rights* 388, 389. Wallis contrasts the development of the Kimberley Process, a more grass-roots model, to the approach by which the United Nations have sought to develop other human rights instruments to regulate international corporations.

4   Wexler, P, *The Kimberley Process Certification Scheme on the Occasion of its Third Anniversary: An Independent Commissioned Review* (Review Report submitted to the Ad Hoc Working Group on the Review of the Kimberley Process, February 2006) 6.

in 2001.[5] The Kimberley Process has 54 participants, representing 81 countries, with the European Union and its 28 member states counting as a single participant.[6]

Noting its growing membership, it can be argued that the Kimberley Process has already made significant progress in reducing the trade in conflict diamonds. Statistics concerning the now monitored legal diamond trade are also encouraging. One indicator of success is the greater number of legitimate diamond exports, demonstrating how the controls have helped reduce illicit trading and bringing more revenues from the trade into the legitimate market. According to a three-year review of the Kimberley Process, the trade in conflict diamonds was less than 0.2 per cent of the overall diamond trade.[7] Sierra Leone is cited as an example of a country where controls have helped maintain the fragile peace, and where the legitimate trade has risen exponentially.[8] In 2008, for example, Sierra Leone exported US$99 million of diamonds, up from $26 million in 2001. Similarly, legitimate rough diamond exports from the DRC were valued at US$552 million in 2008.[9] It should be noted, however, that the black market in diamonds still remains very significant, especially when individual countries are

5   The Role of Diamonds in Fuelling Conflict: Breaking the Link Between the Illicit Transaction of Rough Diamonds and Armed Conflict as a Contribution to Prevention and Settlement of Conflicts, GA Res 55/56, UN GAOR, 55th sess, 79th plen mtg, Un Doc A/RES/55/56 (1 December 2001).
6   Latest figures on participants were taken from the Kimberley Process website, www.kimberleyprocess.com, accessed on 13 August 2015, although the website states that the figures are from 2 August 2013.
7   Ad Hoc Working Group on the Review of the Kimberley Process Certification Scheme, *Kimberley Process Certification Scheme: Third Year Review* (Review Report, Kimberley Process, November 2006) 17; Letter from Karel Kovanda, Kimberley Process Chair, to Kimberley Process Members, 31 January 2007.
8   Global Witness, 'Kimberley Process Certification Scheme Questionnaire for the Review of the Scheme' (Review Submission, 5 April 2006).
9   Kimberley Process Secretariat, 'Annual Global Summary: 2008 Production, Imports, Exports and KPC Counts' (Annual Report Summary, 7 August 2010); Wexler, P, *The Kimberley Process Certification Scheme on the Occasion of its Third Anniversary: An Independent Commissioned Review* (Review Report submitted to the Ad Hoc Working Group on the Review of the Kimberley Process, February 2006).

considered. Despite the progress made in Sierra Leone, for example, it has been estimated that between 10 and 20 per cent of diamonds in that country are still being illegally smuggled, so challenges remain.[10]

The Kimberley Process is distinctive in that it is not a legally binding treaty. While the Kimberley Process Certification Scheme is set out in the form of an international treaty, the instrument is not a legally binding treaty under international law, and is not signed and ratified under standard treaty procedures. However, this less formal structure lends itself to greater flexibility, and wider involvement of non-state actors, which is arguably the strength of the Kimberley Process. Following the informal establishment of the Kimberley Process, commentators such as Global Witness have advocated its being made legally binding through the operation of a UNSC resolution, made with reference to Chapter VII of the UN Charter.[11]

The second distinctive feature of the Kimberley Process is the high level of involvement of businesses and non-governmental organisations in the system.[12] Although not possessing formal voting rights as members, business and NGO groups may be granted observer status and play significant roles in the working groups and committees that comprise the de facto secretariat of the Kimberley Process. For example, in June 2010, the World Diamond Council (WDC), the umbrella

---

10 Interview with Global Witness Representative (telephone interview with author, 30 April 2007). The exact date of the review visit was not specified. In order to obtain information on the operation of the Kimberley Process, a number of standardised questions were presented to government, NGO, and industry participants in the Kimberley Process. The government participant, the Australian Federal Government, provided a written response, while the NGO participant, Global Witness, and the industry participant, Rio Tinto, responded to the questions by means of a verbal interview process. The interviews were recorded and a written transcript produced. In relation to all three participants, the ethical requirements set out by The Australian National University concerning such research were complied with and approved by the ANU Research Ethics Committee.

11 Indications that the KP is not a legally binding treaty are in the Preamble to Kimberley Process, *The Kimberley Process Certification Scheme* (Core Document, 2002), where participants 'recommend' particular measures, the absence of standard signature and ratification clauses that are present in treaty status documents, and the definition of 'Observer' in s I, which includes industry and non-governmental organisation representatives. Curtis has suggested that the Kimberley Process, while not legally binding under international law, is authoritative and is, in fact, a scheme which imposes obligations on its participants. Curtis, K, 'But is it Law?: An Analysis on the Legal Nature of the Kimberley Process Certification Scheme on Conflict Diamonds and its Treatment of Non-State Actors' (2007) Spring *The American University International Law Review*, 26–27.

12 Wright, C, 'Tackling Conflict Diamonds: The Kimberley Process Certification Scheme' (2004) 11(4) *International Peacekeeping* 697, 702.

representative group for the diamond industry, was a member of the monitoring, participation, and statistics committees and was the chair of the diamond experts committee. Global Witness, an NGO with longstanding interest in conflict diamonds issues, was a member of the participation, artisanal and alluvial, statistics, and monitoring committees, while the NGO Partnership Africa Canada was a member of the monitoring, statistics, artisanal and alluvial, and participation committees.[13]

The observer status of industry groups within the Kimberley Process creates direct connections with industry, thereby promoting compliance with the system by the diamond industry, which sees itself as having a stake in the process. The system thus becomes a hybrid of industry self-regulation and government regulation. The institutionalised involvement of NGOs provides further checks and balances to the system. NGOs, possessing independent information and analysis networks, and with links to the international media, provide a scrutineering role that is built in to the Kimberley framework.

Since the emergence of the Kimberley Process, a number of similarly structured organisations have been created, under the general title of multi-stakeholder initiatives. Of particular relevance was the Extractive Industries Transparency Initiative (EITI), which emerged as a result of the World Summit on Sustainable Development held in Johannesburg in 2002. Like the Kimberley Process, the EITI brings together large corporations, governments, and NGOs to achieve its core objectives. The EITI aims to bring financial transparency to the work of extractive industries such as oil, gas, and mining, by requiring corporations to disclose all payments by corporations to governments, and all revenues made by governments from their involvement with these extractive industries. There are 49 corporations that are involved, including large multinationals such as De Beers, ExxonMobil, Shell, and British Petroleum. Also involved with the EITI are nine NGOs,

---

13   Kimberley Process, *The Kimberley Process Certification Scheme* (Core Document, 2002) ss II, VI(8)–(10); Grant, A J and I Taylor, 'Global Governance and Conflict Diamonds: The Kimberley Process and the Quest for Clean Gems' (2004) 93(375) *The Round Table* 385, 386–687.

including Global Witness, Oxfam, Transparency International, and Publish-What-You-Pay. On the governmental level, the EITI has the provisional or full membership of 51 governments.[14]

A preliminary assessment of the EITI suggests that it has made a good start, noting the mild language employed by NGOs, multi-national corporations, and governments indicates a significant level of cooperation. One of the insights noted in relation to the EITI was the relative ease by which it was established, with the EITI relying on a consensus building technique to fully engage with the range of stakeholders. However, the noted drawback of this approach has been the relative difficulty of ensuring effective implementation of the system once it has been created. As implementation mechanisms are often difficult to reach agreement upon, as opposed to general objectives, multi-stakeholder initiatives such as the EITI have begun to experience difficulties when they attempt to bring stakeholders to account in relation to the standards they have agreed to. Behaviour modification is more difficult to address where such processes were not clearly envisaged and set out during the initial negotiations.[15]

Operation of the Kimberley Process

The Kimberley Process functions at two levels, the first being the national level, or the within-country level, where the primary regulator is the national government. At this level, the national

---

14   Statistics about the EITI are located at EITI, *EITI Countries* (2009). Available at: eiti.org/countries. For an analysis of the establishment of the EITI, Koechlin, L and R Calland, 'Standard-setting at the Cutting Edge: An Evidence-based Typology for Multi-stakeholder Initiatives' in A Peters et al. (eds), *Non-State Actors as Standard Setters* (Cambridge University Press, 2009). For further background on multi-stakeholder initiatives in general, see Calton, J M and S L Payne, 'Coping With Paradox Multistakeholder Learning Dialogue as a Pluralist Sensemaking Process for Addressing Messy Problems' (2003) 42(1) *Business and Society* 7; Witte, J M, T Benner and C Streck 'Partnerships and networks in global environmental governance', in U Petschow, J Rosenau and E U von Weizsacker (eds), *Governance and Sustainability: New Challenges for States, Companies and Civil Society* (Greenleaf, 2005); O'Rourke, D, 'Multi-stakeholder Regulation: Privatizing or Socializing Global Labour Standards?' (2006) 34(5) *World Development*; Mena, S and G. Palazzo, 'Input and Output Legitimacy of Multi-stakeholder Initiatives' (2012) 22(3) *Business Ethics Quarterly* 527.
15   Koechlin, L and R Calland, 'Standard-setting at the Cutting Edge: An Evidence-based Typology for Multi-stakeholder Initiatives' in A Peters et al. (eds), *Non-State Actors as Standard Setters* (Cambridge University Press, 2009).

government regulates the national diamond industry by ensuring that all legitimate diamond exports are certified by the means of a Kimberley Process certificate.

The Kimberley Process also operates on the international plane, where the primary regulator is the Kimberley Process Plenary, and the parties that are regulated are the national governments themselves. Important policy decisions are made at the annual plenary meeting of the Kimberley Process, with a consensus mode of decision making: 'Participants are to reach decisions by consensus. In the event that consensus proves to be impossible, the Chair is to conduct consultations.'[16]

The exact meaning of the term 'consensus' has, however, been the subject of divergent opinions. In particular, commentator Ian Smillie points out that some, but not all, government members of the Kimberley Process have considered the word to mean 'unanimity'. Smillie turned to the Oxford Dictionary as the basis for a different meaning for consensus: 'general agreement ... majority view, collective opinion'. Although conceding that other international organisations, such as NATO, equate unanimity with consensus, Smillie presented a different, more sophisticated understanding of the term in his recent paper. He highlighted some failures of the consensus approach as currently practised by the Kimberley Process. In particular, failure to reach consensus occurred as a result of minorities blocking forward movement, disruption, time-wasting, appeasement, lowest common denominator decisions, ineffectual facilitation of critical issues, and lack of confidence and trust. A further challenge occurs where persons are not fully empowered by their representative government or organisation to reach particular negotiated settlements.[17]

Given the challenges faced by a purely consensus approach, Smillie recommends a change to the decision-making procedure to allow voting in the absence of consensus:

16   Kimberley Process, *The Kimberley Process Certification Scheme* (Core Document, 2002) s VI(5).
17   Smillie, I, 'Paddles for Kimberley: An Agenda for Reform' (Report, Partnership Africa Canada, June 2010) 6.

Participants are to reach decisions by consensus. In the absence of consensus, decisions will be made by simple majority of all voting Participants present, except for decisions on those matters specified in Annex A which require a 75% majority of those voting Participants present.[18]

The type of suggestion made by Smillie merits consideration. Even if a consensus decision is viewed as incorporating instances where there are one or two dissenting voices, in contrast to a unanimity approach, it appears necessary and desirable to have an alternative available so that issues can be progressed, following consultation, even where a minority is opposed (i.e. by a majority vote). Smillie suggests that the annex, requiring a 75 per cent majority, be reserved for issues such as additions or deletions from the participants list (i.e. government membership of the Kimberley Process), suspension of participants, and the application of other interim measures relating to non-compliance.[19]

Consensus as a decision-making technique has obvious advantages in terms of optimising the participation of governments and industries in the Kimberley Process. However, in common with other multi-stakeholder initiatives, such as the EITI, this technique held traps for the future evolution of the organisation, where behaviour modification became more important to the future of the organisation than standard-setting. Situations where coercive approaches to behaviour modification are required, by definition, involve departure from unanimity, and put strains on any definition of consensus. In such situations, some type of voting method is preferable.[20]

Between plenaries, the chair plays the central executive role. A vice-chair is elected at each plenary, with the understanding that this representative will assist the chair in this capacity before becoming the new chair in the following year. The chair is also assisted in its executive tasks by a number of committees performing significant regulatory functions. The primary regulatory functions include

---

18   Ibid.
19   Ibid.
20   Koechlin, L and R Calland, 'Standard-setting at the Cutting Edge: An Evidence-based Typology for Multi-stakeholder Initiatives' in A Peters et al. (eds), *Non-State Actors as Standard Setters* (Cambridge University Press, 2009).

annual reporting, peer review, and managing serious non-compliance. Ad hoc working groups have also been formed concerning artisanal mining, diamond technical topics, and the collation of statistics.[21]

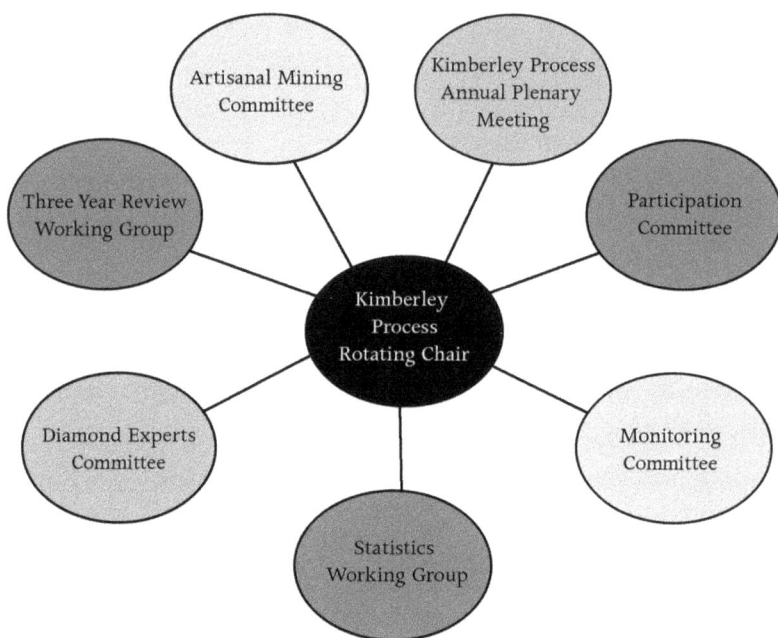

Figure 3.2: Kimberley Process Governance Structures
Source: Author's research.

Accepting New Members

Section VI, paragraph 8 of the Kimberley Process core document states that 'participation in the Certification Scheme is open on a global non-discriminatory basis to all Applicants willing and able to fulfil the requirements of that Scheme'. Apart from the need for a new applicant to provide the chair with 'its relevant laws, regulations, rules, procedures and practices, and update that information

---

21 In its resolution on October 2003, the Kimberley Process Plenary decided to provide a system whereby the vice-chair of the process, elected annually, would automatically become the rotating chair the subsequent year. Kimberley Process, 'Final Communique: Kimberley Process Plenary Meeting' (Sun City, South Africa, 29–31 October 2003). Namibia was the Rotating Chair for 2009. Final Message from Rahul Khullar, Kimberley Process Chair to Kimberley Process Members, 31 December 2008. Israel took the Rotating Chair for 2010. Kimberley Process, 'Kimberley Process Plenary Session Communique' (Swakopmund, Namibia, 5 November 2009) 6.

as required', there were no other stipulations. Despite the simplicity of this requirement, there were initially a number of countries who did not comply. Several had not done so within the first few months of KP operations, and matters had become critical by the end of April 2003 when a special KP plenary meeting was held in Johannesburg. It was agreed at that meeting that a Participation Committee would be struck to examine the credentials of all existing and prospective KP members, to determine whether or not they could meet the minimum standards. It was agreed that there would be a 'tolerance period' until 31 May 2003, during which all participants and prospective participants would submit information relevant to their membership. The tolerance period was extended to June, then to the end of July, and finally, with a chair's notice at the end of July, to August 31.[22]

The Participation Committee included seven governments (Angola, Canada, the European Community as it was then called, Israel, Russian Federation, South Africa, and the United States), NGOs (Global Witness and Partnership Africa Canada) and the WDC. During this period, the committee examined the legislation, regulations, and relevant documentation of every participant.[23]

At this time a euphemism for removing a country from the KP was developed, which would be 'dropped from the list'. Following the examination of credentials, several countries were dropped from the list: Brazil, Burkina Faso, Cyprus, Gabon, Malta, Mexico, Norway, Philippines, and Poland. Three of these countries — Brazil, Mexico and Norway — subsequently rejoined the KP.[24]

The Participation Committee play a central role in the Kimberley Process international enforcement system. The Kimberley Process core document states that participation in the certification scheme is open on a global, non-discriminatory basis to all applicants willing and able to fulfil the requirements of the scheme. It is significant that only those applicants willing and able to fulfil the requirements of the

---

22   Smillie, I, 'Paddles for Kimberley: An Agenda for Reform' (Report, Partnership Africa Canada, June 2010) 6–7.

23   Ibid.

24   Ibid.

scheme will be able to join. The work of the Participation Committee is largely concerned with whether prospective members are able to fulfil these requirements, such that they may be admitted to the scheme.[25]

The Participation Committee Terms of Reference elaborates on the functioning of that committee.[26] Like other committees, the Participation Committee must include representatives from NGOs and industry, as well as governmental representatives. The chair must also ensure an appropriate geographical balance, and that there is appropriate expertise on the committee to perform its functions.

The Participation Committee is tasked with assisting the chair in its role of handling the admission of new applicants to the Kimberley Process, and may enter into a dialogue with the applicant on issues to be addressed.[27]

Annual Reporting

The Kimberley Process operates internationally to ensure compliance by national governments. Participating governments are required to provide information on an annual basis on the way in which they are implementing the requirements of the Kimberley Process.[28] Annual reports must include information on the national laws and regulations for the export and import of rough diamonds; internal controls prior to the export of diamonds and after import; penalties for individuals and companies contravening diamond regulations; the collection of import and export data; whether there is a procedure for issuing Kimberley Process certificates; whether the certificate fulfils security requirements; evidence to be provided by exporter as proof diamonds as not conflict tainted; and the number of Kimberley Process certificates issued and to which participating governments they were sent.[29]

---

25   Kimberley Process, *The Kimberley Process Certification Scheme* (Core Document, 2002) s VI (8).
26   Kimberley Process, *Administrative Decision: Participation Committee Terms of Reference* (Plenary Meeting Decision, Gatineau, Quebec, 29 October 2004).
27   Kimberley Process, *The Kimberley Process Certification Scheme* (Core Document, 2002) ss II, V(a), VI(8), (9).
28   Kimberley Process, *The Kimberley Process Certification Scheme* (Core Document, 2002) s VI (11); Kimberley Process, *Administrative Decision: Implementation of Peer Review in the Kimberley Process* (Plenary Meeting Decision, Sun City, South Africa, 30 October 2003).
29   Administrative Decision on Annual Reporting, Annex I.

Also to be included in annual reports is information on the system of internal controls and industry self-regulation that is implemented in the participant state, information about statistical collection, and observations on experiences, problems and solutions that have been noted during the implementation process.

The collection of statistics in an area of economic regulation such as the international diamond trade is naturally a central aspect of effective management. Such data is particularly important for identifying any irregularities or anomalies that could indicate that conflict diamonds are entering the legitimate trade. Kimberley member governments are required to keep quarterly aggregate statistics on rough diamond exports and imports in a standardised format, as well as numbers of certificates validated for export, and imported shipments accompanied by certificates. Annual statistics in the rough diamond trade, listed according to country, are now publicly available through a dedicated website maintained by a Kimberley Process Working Group on Statistics. Statistics on exports and imports must record diamond origin and provenance, carat weight, and value.[30]

The Working Group on Statistics is mandated to deal with statistical matters pertaining to rough diamonds, particularly in respect to the production and trade in rough diamonds, to ensure the effective implementation of the Kimberley Process.[31] Like the Monitoring Committee, the Statistics Committee's role includes general policy development in the area of statistics, including the use of common classification systems. Its second role is concerned with statistical collation and analysis, and administrative support to the Kimberley Process. Should a member government fail to provide statistics within three months of the close of a quarter, then issue of the continued membership of that government will be forwarded to the Participation Committee for consideration and possible compliance action.[32]

In early KP negotiations, statistical data was regarded by some governments as information that could not be shared, either internally among KP participants, or externally. Some countries cited commercial

---

30   Kimberley Process, *The Kimberley Process Certification Scheme* (Core Document, 2002) Annex III. The document refers to Harmonised Commodity Description and Coding System (HS) classifications 7102.10, 7102.21, and 7102.31.

31   Kimberley Process, *Working Group on Statistics Terms of Reference* (29 April 2003).

32   Kimberley Process, *Administrative Decision on Statistical Reporting* (undated).

sensitivity as a reason Russia treated diamond production data as a state secret, and said that it would not go along with a certification system that would reveal this secret to others.[33]

By 2003, however, much of the sensitivity on statistics had diminished, and in 2004 even Russia had agreed to submit quarterly trade data and semi-annual production data. The KP statistics website is today the best source of data on rough diamond production and trade, and is an essential tool in tracking anomalies in the system.[34]

For several of its early years, however, the KP statistics website was accessible to participants only. There was very strong resistance to making any of the data public, with governments citing 'commercial sensitivity'. Nevertheless, in the past two years greater, although not complete, statistical openness has been achieved, without any apparent ill effect. The major advantage appears to be an end to charges that the Kimberley Process was hiding something by refusing to make its statistics public.[35]

The Working Group on Monitoring is tasked with reviewing annual reports by member governments and reporting to the plenary. In doing so, the Working Group on Monitoring must draw on available statistical data, and work cooperatively with the Statistics Working Group and the Participation Committee. Other Kimberley members may also present reports for consideration by the Working Group on Monitoring.[36]

---

33   Smillie, I, 'Paddles for Kimberley: An Agenda for Reform' (Report, Partnership Africa Canada, June 2010) 13.

34   Ibid.

35   Ibid.

36   The selection of office-holders to chair working groups, ad hoc bodies, etc. is to be decided by participants in plenary following consultation by the chairperson. The terms of reference for the Monitoring Committee were approved at the 28–30 April 2003 plenary, but were revised at the Gaborone Plenary meeting in November 2006 as a result of the three-year review of the KP. Kimberley Process, *Administrative Decision about the KPCS Peer Review System* (Plenary Meeting Decision, Gaborone, Botswana, November 2006).

The Participation Committee must also be informed by the Working Group on Monitoring, via the chair, of government members that have failed to submit an annual report from the previous year and countries that have failed to provide the required statistical data.[37]

Peer Review: General Operation

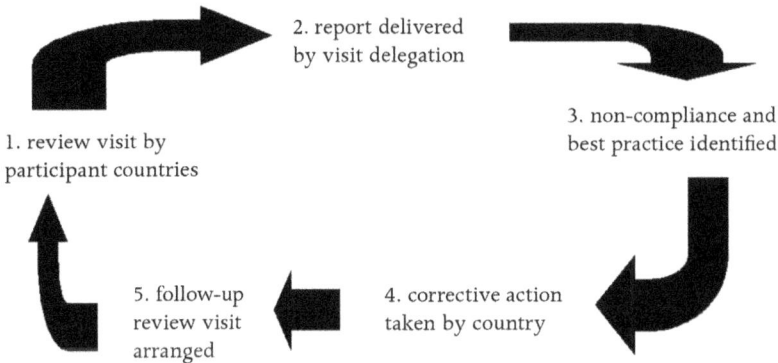

2. report delivered by visit delegation

3. non-compliance and best practice identified

1. review visit by participant countries

5. follow-up review visit arranged

4. corrective action taken by country

Figure 3.3: Kimberley Process Peer Review
Source: Author's research.

Monitoring is essential in any system dealing with standards and supply chains. A number of global commodity governance systems have evolved over time to include rigorous and credible third-party verification systems, such as the Forest Stewardship Council, the Fair Labour Association, and the Responsible Jewellery Council. From the beginning, monitoring was a highly contentious subject in Kimberley Process negotiations. Diamonds were regarded as a strategic mineral in Russia, for example, and data regarding production and trade was classified. In many countries there were commercial sensitivities and

---

37   Kimberley Process, *The Kimberley Process Certification Scheme* (Core Document, 2002) s V. A challenge that has been addressed by the rotating chair relates to the appearance of fraudulent KP certificates in Sierra Leone and Ghana. Following the appearance of such certificates, particular certificates were removed from certification and brought to the attention of other KP participants. Kovanda, K, Kimberley Process Chair, 'The Appearance of Fraudulent Certificates' (Letter to Kimberley Process Members, 23 March 2007).

security issues. In the initial KP agreement, there was provision only for monitoring in cases of 'significant non-compliance', a term that was never defined.[38]

It is also arguable that the flexibility of the Kimberley Process, with its focus on diplomacy, consensus, and information sharing, has emphasised the autonomy and sovereign equality of state actors, thus constituting a horizontal approach to regulation, rather than focusing solely on the vertical dimension, involving the cession of authority to a supranational body. The horizontal approach is particularly evident in relation to the system of monitoring through the peer review system. In contrast with other international systems that focus on adversarial dispute resolution before an international tribunal, the focus on peer review is a novel approach. It arguably engages more effectively with government members by promoting a sense of ownership for the certification system. States must take responsibility for the effective functioning of the system, rather than ceding this role to a supranational entity.[39]

The Monitoring Committee is responsible for the implementation of the peer-review process within the Kimberley Process. Peer review involves two central tasks: monitoring the implementation of the Kimberley Process by participant countries through review visits,[40] and monitoring situations of serious non-compliance through review missions.[41]

Although review missions and visits are conducted with the consent of the government member, a review team must be given the full cooperation of the authorities of the country under review, who should facilitate access to governmental institutions and organisations

---

38   Smillie, I, 'Paddles for Kimberley: An Agenda for Reform' (Report, Partnership Africa Canada, June 2010) 9.

39   Other international organisations have monitoring mechanisms that function in similar ways to review missions, such as investigative missions for the United Nations High Commissioner for Human Rights. However, these missions are typically carried out at the supranational level, rather than being staffed by representatives of member nations themselves, as occurs within the Kimberley Process.

40   Kimberley Process, *Administrative Decision: Implementation of Peer Review in the Kimberley Process* (Plenary Meeting Decision, Sun City, South Africa, 30 October 2003); Kimberley Process, *The Kimberley Process Certification Scheme* (Core Document, 2002) s VI.

41   Kimberley Process, *Administrative Decision: Implementation of Peer Review in the Kimberley Process* (Plenary Meeting Decision, Sun City, South Africa, 30 October 2003).

relevant to the implementation of the Kimberley Process, and solicit the cooperation of industry, consistent with national law and organisational rules and regulations.[42]

Review missions and visits should generally number five members in total, consisting of three government members, an observer from civil society, and an observer from the private sector. The Administrative Decision on Peer Review further provides that, in nominating government member representatives, the chair would seek to ensure geographical balance and adequate balance between countries that are primarily engaged in production, trading, and processing of rough diamonds.[43] They should last between two and five working days, with dates determined with the consent of the government member being reviewed. The fact that individual members of the review mission must provide their own expenses presents a challenge for civil society representatives, particularly those representing artisanal miners from diamond-producing nations. To date there has been some limited sponsorship made available from government members to support the important role of civil society experts in these teams.[44]

The leader of the review mission or visit must draw up a written draft report giving an account of the activities of the mission and its findings, in particular reflecting the implementation of the Kimberley Process in the reviewed country. The draft report must be submitted simultaneously to the chair and the reviewed country.

The government under review has a right of reply in relation to the review report, and may send observations to the chair and members of the review mission or visit within a month of the submission of the draft report. The chair may then invite the government's authorities to discuss the observations with the review team members to clarify any misunderstandings. In the event that disagreement persists, the report

---

42   Ibid.
43   Ad Hoc Working Group on the Review of the Kimberley Process Certification Scheme, *Kimberley Process Certification Scheme: Third Year Review* (Review Report, Kimberley Process, November 2006) 41.
44   Kimberley Process, *Administrative Decision: Implementation of Peer Review in the Kimberley Process* (Plenary Meeting Decision, Sun City, South Africa, 30 October 2003).

will be circulated along with the observations of the government member to other countries and observers. The chair may add its own observations as well.[45]

Review visit and review mission reports are considered confidential, as between government and non-government members within the Kimberley Process system. The administrative decision also provides for follow-up action in relation to a review mission or review visit. Where the review mission deems it necessary and appropriate, the chair may recommend to the plenary the sending of a follow-up mission or review visit.[46]

The Sun City Plenary in 2002 provided for a roster of experts to be drawn up by the chair on a recommendation from the Monitoring Committee. Since then, a substantial number of government and non-government Kimberley members have nominated experts for inclusion in the roster that, as of July 2006, comprised 97 experts representing government, industry, and civil society. On the basis of this very positive response, it was possible in all cases for teams to be appointed corresponding to the required criteria. To date, experts from 17 different government members, and from all major geographical regions represented in the Kimberley Process, have participated in review visits.[47]

It was possible for a number of experts from developing, artisanal-alluvial producing countries to participate in review visits and review missions. Participation by such experts is of great importance because of the crucial role of the visits in disseminating best practices among government members and teaching participating experts. The KP three-year review recommended that the participation of experts from artisanal-alluvial–producing nations in as many review visits as possible should be continued and, if possible, further developed, above all in review visits to artisanal-alluvial–producing countries.[48]

---

45   Ibid.
46   Kimberley Process, *The Kimberley Process Certification Scheme* (Core Document, 2002) s VI (16); Kimberley Process, *Administrative Decision: Implementation of Peer Review in the Kimberley Process* (Plenary Meeting Decision, Sun City, South Africa, 30 October 2003).
47   Ad Hoc Working Group on the Review of the Kimberley Process Certification Scheme, *Kimberley Process Certification Scheme: Third Year Review* (Review Report, Kimberley Process, November 2006) 41.
48   Ibid.

The Participation Committee also has a role in relation to compliance. It considers information submitted to it by the Working Group on Monitoring regarding compliance by a government member, and can determine whether the government is able and willing to meet the minimum common standards of the certification scheme.[49]

Peer Review: Cases of Serious Non-Compliance

Through the action of the peer-review system, in tandem with the working groups on monitoring and participation, the Kimberley Process has responsive yet effective mechanisms for holding member governments to account regarding their obligations under the system. The Kimberley Process commands the threat of a significant sanction, which is the expulsion of members for non-compliance, through which the diamond trade with that member is prohibited to other Kimberley Process participant countries. Through sharing of information and experience, with heavy reliance on consensus and diplomatic pressure, the Kimberley Process offers a novel approach for addressing urgent global issues.[50] The informality of its mode of operation has, arguably, enhanced its ability to respond quickly to crises that have occurred in the system, notably in the Republic of Congo-Brazzaville (RCB) and Côte d'Ivoire.[51]

Two straightforward cases of apparently serious non-compliance occurred early in the life of the Kimberley Process. In May 2003, the Central African Republic (CAR) was suspended from the KP following a coup in which François Bozizé overthrew the government of President Ange-Félix Patassé and suspended the constitution. The CAR was reinstated as a participant after authorities provided

---

49 Kimberley Process, *Administrative Decision: Participation Committee Terms of Reference* (Plenary Meeting Decision, Gatineau, Quebec, 29 October 2004) s 4.1; Kimberley Process, *The Kimberley Process Certification Scheme* (Core Document, 2002) s VI (8).

50 Wexler, P, *The Kimberley Process Certification Scheme on the Occasion of its Third Anniversary: An Independent Commissioned Review* (Review Report submitted to the Ad Hoc Working Group on the Review of the Kimberley Process, February 2006) 6.

51 Subsequent to the action taken by the Kimberley Process Chair in relation to the Côte d'Ivoire situation, the Kimberley Process Plenary issued resolution and follow-up action, including cooperation with the United Nations. Kimberley Process, 'Final Communique: Kimberley Process Plenary Meeting' (Moscow, Russia, 15–17 November 2005). Other participants have also been suspended from the Kimberley Process for differing periods of time. Ghana was suspended following an Administrative Decision of the Gaborone Plenary on 9 November 2006, but was reinstated by the KP chair on 1 March 2007. Letter from Karel Kovanda, Kimberley Process Chair, to Kimberley Process Members, 1 March 2007.

assurances they could implement the KP and agreed to let a review mission evaluate the country's diamond control system. The review found that CAR was managing its internal diamond controls and KP standards responsibly.[52]

Lebanon expressed its eagerness to join the Kimberley Process in early 2003. It submitted all of the required documentation, including legislation that at the end of the tolerance period was awaiting presidential signature. Lebanon was included in the list comprising 39 countries plus the European Community that was approved with effect from 31 August 2003. However, nine months later, the presidential approval for the country's KP legislation had not been given and Lebanon was dropped from the list on 1 April 2004. In 2005, Lebanon was readmitted to the Kimberley Process following enactment of the legislation and two KP review missions.[53]

The first major test of the ability of the Kimberley Process to manage serious non-compliance was in relation to the RCB, which neighbours the DRC. Following consideration of relevant statistics and reports, it was brought to the attention of the Kimberley Process that the RCB appeared to be funnelling diamonds mined in the DRC through its borders, thereby bypassing the certification requirements that would otherwise have been implemented in the DRC. The chair of the Kimberley Process took rapid action, firstly securing the agreement of the RCB President that a KP review was required, and authorising the deployment of a review mission that verified the problem. The review took place in May 2004, and included an aerial survey of the country's diamond mining areas. The review concluded that the RCB's exports could not be explained by local production or official imports.[54]

In July, through a chair's notice, RCB was expelled from the Kimberley Process, meaning that other countries no longer traded in diamonds with that country. Conditions for readmission included an independent third-party survey of the country's geological diamond potential. The result of the expulsion was that the legitimate, certified

---

52   Smillie, I, 'Paddles for Kimberley: An Agenda for Reform' (Report, Partnership Africa Canada, June 2010) 6–7.

53   Ibid.

54   Ibid 7; Global Witness, 'Kimberley Process Certification Scheme Questionnaire for the Review of the Scheme' (Review Submission, 5 April 2006); Interview with Global Witness Representative (telephone interview, 21 May 2007).

trade through the DRC picked up significantly. Three years later, in November 2007, RCB hosted another KP review, which concluded that it had met all of the Kimberley Process stipulations. The RCB was then readmitted into the Kimberley Process Certification Scheme.[55]

In some respects, the Côte d'Ivoire situation is a good example of the Kimberley Process acting quickly to respond to a serious threat to its integrity. The Kimberley Process was able to rapidly make a decision to prohibit the trade in diamonds originating from Côte d'Ivoire, thereby providing at least a formal barrier to conflict diamonds. The Kimberley Process, through its Working Group on Monitoring, was apprised of the situation in Côte d'Ivoire in September 2004, following the emergence of indications that diamond production was continuing in rebel-controlled northern Côte d'Ivoire. The Working Group on Monitoring adopted a recommendation to the chair on the matter on 24 September 2004, after which the chair entered into communications with the Côte d'Ivoire authorities, who clarified that an export ban had been imposed on rough diamond throughout Côte d'Ivoire.[56] The Chair of the Kimberley Process issued a request on 23 November 2004 to all participants not to accept any shipments with Côte d'Ivoire certificates until further notice, and reported on the status of Côte d'Ivoire at the Ottawa Plenary in October 2004.[57] It should be noted, however, that UNSC expert panels noted ongoing diamond mining in Côte d'Ivoire during the civil war, even after the country was excluded from the Kimberley Process.[58]

---

55 Global Witness, 'Kimberley Process Certification Scheme Questionnaire for the Review of the Scheme' (Review Submission, 5 April 2006); Interview between Author and Global Witness Representative; Smillie, I, 'Paddles for Kimberley: An Agenda for Reform' (Report, Partnership Africa Canada, June 2010) 7–8. However, some commentators have questioned whether expulsion of the Republic of Congo would be sufficient to stop the conflict diamonds trade in the neighbouring DRC. Malamut, S A, 'A Band-Aid on a Machete Wound: The Failures of the Kimberley Process and Diamond-Caused Bloodshed in the Democratic Republic of Congo' (2005) 29 *Suffolk Transnational Law Review* 25, 26, 44–45.
56 Kimberley Process Working Group on Monitoring, 'Submission for the 2006 Review of the KPCS' (Kimberley Process Secretariat, February 2006) 19.
57 Ibid 19–20.
58 Update Report of the Group of Experts Submitted Pursuant to Paragraph 2 of Security Council Resolution 1632 (2005) Concerning Côte d'Ivoire, UN Doc S/2006/204 (31 March 2006).

Other countries in which serious compliance issues have arisen include Zimbabwe, Ghana, Bangladesh, Brazil, and Venezuela.[59] In mid-2005, Venezuela, a KP participant since 2003, ceased issuing Kimberley Process certificates, and communications with the KP ceased. Nevertheless, diamonds were being mined and openly, if not legally, exported. The problems were documented in a 2006 Partnership Africa Canada report. The KP procrastinated, and it was not until October 2008, following bitter internal debate and widespread calls for Venezuela's expulsion from the KP, that a KP team visited Venezuela, corroborating many of Partnership Africa Canada's findings. In November 2008, Venezuela announced that it would self-suspend from the KP, saying it would halt all diamond production and trade for at least two years while reorganising its diamond sector. The KP concurred with this approach.[60]

However, in Venezuela little changed. Early in 2009, the mineral leases of five diamond mining cooperatives held by the state-owned mining concern Corporación Venezolana de Guayana were renewed. Diamond mining and exporting, whether legal or illegal, continued as before.[61]

In a March 2009 letter, the KP Chair, Namibia, hailed the arrangement with Venezuela, saying that the KP would 'assist and support the country in developing appropriate internal controls over its alluvial diamond mining'. The chair said that this was 'yet another example of mutual inclusiveness inherent in the Scheme and is testimony to the willingness of the KP family to stand together, learn from global best practices and proactively provide assistance when required'.[62]

However, there has been little substantive communication with Venezuela since 2008, and the KP has provided no assistance or support, but rather turned a blind eye to the fact that Venezuela's diamonds are entering world markets illegally.[63]

---

59   Letter Prior to Plenary Meeting in Brussels from Karel Kovanda, Kimberley Process Chair to Kimberley Process Members, 2007.
60   Smillie, I, 'Paddles for Kimberley: An Agenda for Reform' (Report, Partnership Africa Canada, June 2010) 8.
61   Ibid.
62   Ibid.
63   Ibid.

In 2008, a number of events occurred suggesting that Zimbabwe was losing its ability to meet KP minimum standards. Large volumes of easily identified smuggled Zimbabwean diamonds were the subject of arrests in Dubai and India. A diamond rush by illicit diamond diggers in the Marange area was suppressed by well-documented extrajudicial killings and widespread human rights abuse by the buying offices in Manica, just across the Zimbabwe border in Mozambique, where a flourishing trade in smuggled goods continues today.[64]

Zimbabwe became the subject of bitter debate within the KP. It took months before a review mission could be undertaken, and the mission itself became the subject of debate and political manipulation. Its findings were clear but its recommendations were vague, and in the end a bitter debate resulted in little more than the appointment of a KP monitor whose terms of reference omitted almost all the topics of controversy. Zimbabwe's continued presence without censure in the KP has been ensured by strong support from South Africa and other neighbouring countries, although its behaviour continues to be both erratic and controversial, and its ability to meet minimum KP standards cannot be demonstrated.[65]

The Zimbabwean situation has arguably been the most concerning, with reports of violence and smuggling in the Marange mining area following the occurrence of a diamond rush there. As a result, the Kimberley Process agreed to send a review visit to the country. Although irregularities were discovered, the Kimberley Process decided against suspending Zimbabwe's membership in the scheme. However, a Joint Work Plan was agreed between the Kimberley Process and Zimbabwe, involving the appointment of a special Kimberley monitor to address compliance issues. A 'supervised export mechanism' was established for exports of rough diamonds from Marange, under which the Kimberley Monitor must examine potential exports and sign the Kimberley Process certificate before they can be

---

64   Ibid 8–9.
65   Ibid.

considered as legitimate.[66] One of the consequences of the Marange diamonds dispute within the Kimberley Process, when it became clear that Zimbabwe would be permitted to export diamonds from Marange, was that NGO members from the KP, including Global Witness, chose to leave the KP completely, taking the view that it had failed in its core mandate.[67] Other NGOs, such as Partnership Africa Canada, have chosen to remain within the KP despite their objections to the handling of the issue.[68] With the Marange diamonds issue settled in its favour, Zimbabwe has gone on to export a large quantity of diamonds legitimated by the Kimberley Process: it exported US$480 million of rough diamonds in 2014.[69]

Review missions have an important role where there are more serious allegations of non-compliance. The case of Brazil is a good example. In 2005, Partnership Africa Canada published a report detailing instances of fraud relating to illicit diamond production and exports within Brazil's diamond industry. This led Brazil's Federal Police to launch an investigation into the country's diamond industry, and Brazilian authorities to suspend diamond exports. A previously commissioned Kimberley review visit found a range of shortcomings in the way in which the scheme was implemented in Brazil, ranging from a lack of training and experience among staff and customs officers, to flaws

---

66   Esau, Bernhard, MP, Kimberley Process Chair, 'Public Statement on the Situation in the Marange Diamond Fields, Zimbabwe' (Kimberley Process Secretariat, Windhoek, Namibia, 26 March 2009); Esau, Bernhard, MP, Kimberley Process Chair, 'Statement: High Level Envoy Visit to Zimbabwe: Situation in Marange Diamond Fields' (Kimberley Process Secretariat, Windhoek, Namibia, 16 April 2009); Esau, Bernhard, MP, Kimberley Process Chair, 'Risk of Fake KP Certificates' (Kimberley Process Secretariat, Windhoek, Namibia, June 2009); Kimberley Process Secretariat, 'Kimberley Process Intersessional Meeting Communique' (Windhoek, Namibia, 25 June 2009); Esau, Bernhard, Hon, Kimberley Process Chair, 'Clarification About the Kimberley Process Chair's Working Visit to Zimbabwe Which Took Place 19–21 August 2009' (Kimberley Process Secretariat, Windhoek, Namibia, 3 September 2009); Hirsch, Boaz, Kimberley Process Chair, 'Appointment of KP Monitor to Zimbabwe' (Kimberley Process Secretariat, Jerusalem, Israel, 1 March 2010); Hirsch, Boaz, Kimberley Process Chair, 'Re: Trade of Marange Diamonds in Compliance with KPCS Requirements: Vigilance Against the Laundering of Illicit Shipments' (Letter to Kimberley Process Participants, 6 May 2010).
67   Global Witness, 'Global Witness Leaves Kimberley Process, Calls for Diamond Trade to be Held Accountable' (2 December 2011). Available at: www.globalwitness.org/archive/global-witness-leaves-kimberley-process-calls-diamond-trade-be-held-accountable/.
68   Partnership Africa Canada, 'PAC and the Kimberley Process: A History', www.pacweb.org/en/pac-and-the-kimberly-process; a listing of remaining members of the official KP civil society observers is available at the Kimberley Process website: www.kimberleyprocess.com/en/civil-society-coalition.
69   Kimberley Process website: www.kimberleyprocess.com/node/227.

in Brazil's Kimberley certificate.[70] Brazil handled the revelation of irregularities well. On its own accord, Brazil suspended all diamond exports for 2006 well in advance of the Kimberley review visit. Brazilian authorities were transparent and eager to assist the review team. Brazil invited a follow-up review visit.[71]

Peer Review: Evaluation

In the initial KP agreement, there was provision only for monitoring in cases of 'significant non-compliance', a term that was never defined. A year after the KP came on stream in 2003, the peer-review system was agreed to. The entire process was developed after the core document had been finalised, providing a textbook case of how the KP has evolved in order to achieve the goals that are expected of it.[72]

The peer-review mechanism was assessed as part of the November 2006 review of the Kimberley Process, which was conducted by an Ad Hoc Working Group, as envisaged in the Kimberley core document. The group reported that there was widespread agreement among governments, civil society, and NGOs that the peer-review mechanism had been a great success. In particular, the review highlighted the fact that review visits were a crucial confidence-building tool, allowing the Kimberley Process to be sure its requirements were being met effectively.[73] The review noted that 32 participants received review visits prior to the review, and that a further two non-participants, Liberia and Lebanon, had received special expert missions prior to joining the Kimberley Process. Two review missions had been carried out: to the Central African Republic and the Republic of Congo. As at 30 October 2006, 12 member governments had not yet received review

70   Ad Hoc Working Group on the Review of the Kimberley Process Certification Scheme, *Kimberley Process Certification Scheme: Third Year Review* (Review Report, Kimberley Process, November 2006) 39–40.
71   Ibid 40.
72   Smillie, I, 'Paddles for Kimberley: An Agenda for Reform' (Report, Partnership Africa Canada, June 2010) 8–9.
73   Kimberley Process, *The Kimberley Process Certification Scheme* (Core Document, 2002) s 20; Ad Hoc Working Group on the Review of the Kimberley Process Certification Scheme, *Kimberley Process Certification Scheme: Third Year Review* (Review Report, Kimberley Process, November 2006). See also Ad Hoc Working Group on the Review of the KPCS, *Kimberley Process Certification Scheme Questionnaire for the Review of the Scheme* (Questions for Review Submissions, Kimberley Process, 2005); Kimberley Process, *Administrative Decision: Terms of Reference Ad-Hoc Working Group on the Review of the KPCS* (Revised 31 July 2006).

visits, but nine invited review visits. Overall, 42 of the 45 Kimberley participants (93 per cent) received or invited review visits or missions. All previously conflict diamond–affected countries received review visits, and almost all countries reporting diamond production or trade had had review visits.[74]

The review, however, also mentioned a number of areas in which the peer-review mechanism could be improved. Croatia, Indonesia, and Venezuela had not had review visits, and had not requested them. The three-year review recognised that these three countries should be encouraged as strongly as possible to invite review visits as soon as possible. The chair of the Monitoring Committee confirmed that these countries had been approached and are considering inviting a review visit.[75]

The three-year review also recommended a number of modifications to the Monitoring Committee mandate, specifically so that review visit activities could explicitly integrate a regional dimension into their activities — i.e. trading patterns in neighbouring countries. Expert missions should be able to be deployed on an ad hoc basis in preparation for determining whether or not to admit a particular applicant into the Kimberley Process.[76]

The three-year review considered the effectiveness of review visits in two ways: by considering whether the system had detected the main implementation problems, and whether the peer-review system had contributed to bringing about tangible improvements to the identified problems. The Review Working Group concluded that governments who had been reviewed had received the visits with openness, and provided access to their documentation and the range of activities linked to certification. This made it possible for the review team to identify issues that would not otherwise have been apparent.

However, the three-year review also noted that some review visits were more sophisticated in comparing the actual mining capacity of a reviewed government to its declared rough diamond exports.

---

74   Ad Hoc Working Group on the Review of the Kimberley Process Certification Scheme, *Kimberley Process Certification Scheme: Third Year Review* (Review Report, Kimberley Process, November 2006) 38.
75   Ibid.
76   Ibid.

The review recommended that each team report on whether it has reviewed internal controls for effective compliance in the countries it has visited. The three-year review also suggested that individual countries identify different needs for technical assistance and training in order to help participating governments implement effective internal controls. Examples of constructive interaction, resulting in improved implementation practices, included the adoption of proper import procedures by some producing governments that did not have import procedures in place; the training of diamond valuators to enable them to carry out a review visit's recommendation that all imports into a government member be subjected to a regime of physical inspection; and the initiation of an investigation by a government into suspicious trading activities pursuant to the findings of a review visit. One suggestion to improve the capacity of review visits was to increase the length of the average review visit. Given that these are normally only three to five days long, they can take on more of a diplomatic than an investigative character.[77]

The review also recommended that the Kimberley Process should seek to further diversify the leadership of review visits to include in particular alluvial-producing countries, as only Sierra Leone in 2006 had taken such a role. It was also suggested that the criteria in the Administrative Decision on Peer Review should be expanded to include a provision that experts are required to be impartial and highly professional, and should further require members to disclose any potential conflict of interest.[78]

The review suggested that the peer-review system be maintained, but that the Administrative Decision on Peer Review be amended, specifying that, in further review visits, attention should be focused on follow-up of issues identified in the first visit. In the case of repeated review visits, the visiting teams should be flexible in size and duration, to ensure that scarce resources are focused on substantial implementation issues. The Participation Committee

---

77   Ibid 38–39; Interview between Author and Global Witness Representative.
78   Ad Hoc Working Group on the Review of the Kimberley Process Certification Scheme, *Kimberley Process Certification Scheme: Third Year Review* (Review Report, Kimberley Process, November 2006) 41-42.

should carefully engage with countries that fail to implement the review visit recommendations, with expulsion from the KP available as a last resort.[79]

One of the areas in which the peer-review could be strengthened would be to make the receipt of a peer-review visit compulsory to all members. The reluctance of some members to receive a peer-review visit calls into question the entire monitoring and enforcement system of the Kimberley Process. If a country is able to deny access to a review team, there is no ability for the Kimberley Process to verify its level of compliance with the process.

In evaluating the Kimberley Process's response to cases of serious non-compliance, its ability to take executive action between plenary meetings, mediated by the chair, has been very important. In this feature, a year-round operational secretariat, it resembles the success of the equivalent secretariat responsible for managing compliance with the *Convention on the Illegal Trade in Endangered Species of Wild Flora and Fauna*. A system has been developed for the convention whereby a decision concerning a serious non-compliance issue can be made during the course of a year, prior to a meeting of the plenary. Action such as a trading ban with the problematic country can be implemented in this interim period.[80]

Although the office of the Kimberley Process Chair has been highlighted as a strength of the peer-review system, it has also been criticised for being overly dependent on the willingness of the incumbent to take decisive action. For example, observers have argued that it was fortunate that the Canadians were chairing the Kimberley Process at the time that the Republic of Congo was expelled, as they were proactive and took the decision to ensure the integrity of the system.[81] This example can be contrasted with the situation of Venezuela. Although seen as being in blatant non-compliance with Kimberley requirements, Venezuela was not targeted for expulsion under the mandate of the European Community (2007)

79   Ibid 46; Interview between Author and Global Witness Representative.
80   Hewitt, T, 'Implementation and Enforcement of the Convention on International Trade in Endangered Species of Wild Fauna and Flora in the South Pacific Region: Management and Scientific Authorities' (2002) 2(1) *Queensland University of Technology Law Journal* 98, 98–130.
81   Interview between Author and Global Witness Representative.

or Angola (2008).[82] This highlights the need for a formal procedure for dealing with countries that are seriously non-compliant, and, arguably, in determining the suitability of particular governments to carry out this important function.[83]

When it works well, the peer-review system is adequate, although three-day reviews in some cases are not long enough to develop a comprehensive understanding of a country's diamond industry. In many cases, however, it is far from adequate. Worst case examples include a review of Ghana where the report, a year in production, was superseded by a much tougher UN report revealing the transit through Ghana of conflict diamonds from Côte d'Ivoire (missed entirely by the KP team). An enormous nine-member Guinea review team spent less than two hours outside the capital city and did not complete its report for more than a year. A review of Venezuela was orchestrated entirely by the non-compliant host government. Civil society was prevented from participating in the exercise, and the team was never allowed near diamond mining or trading areas.[84]

The makeup of review teams is inconsistent. Burden sharing has been uneven, with some NGOs footing a larger share of review costs than most governments. This has been alleviated in recent years by contributions from Rio Tinto Diamonds, Norway, Switzerland, and the United States to a fund for NGO participation.[85]

82   Ibid; Interview between Author and Global Witness Representative. Subsequently, an Administrative Decision (AD) on Venezuela's participation was adopted in the plenary meeting of the Kimberley Process in Brussels in 2007. Further to this, the KP Chair attempted to organise a review visit in the first quarter of 2008 but this was never acceded to by Venezuela. As a result, the Working Group on Monitoring concluded on 10 June 2008 that, since it could not ensure the implementation of the administrative decision, that the matter should be referred to the Participation Committee. Just prior to the intersessional meeting, the KP Chair received a notification from Venezuela that it intended to 'voluntarily separate from the KP for a period of two years and to cease certification for export of its diamonds'. Message regarding Compliance of Venezuela from Kimberley Process Chair Rahul Khullar to Kimberley Process Members, 9 July 2008. The plenary formally encouraged continued efforts to reintegrate Venezuela into the KP at its meeting in 2009. Kimberley Process, 'Kimberley Process Plenary Session Communique' (Swakopmund, Namibia, 5 November 2009).
83   Interview between Author and Global Witness Representative.
84   Smillie, I, 'Paddles for Kimberley: An Agenda for Reform' (Report, Partnership Africa Canada, June 2010) 10.
85   Ibid.

While some reviews have been thorough and have made important recommendations, there has been a chronic lack of follow-up. Review teams have repeatedly stated that some of the countries worst affected by conflict diamonds, such as Angola, the DRC, and Sierra Leone, have extremely weak internal controls. Getting a grip on internal controls remains the single most important issue for the diamond industry and the Kimberley Process.[86]

Well-documented cases of serious non-compliance have been brought to the attention of the Kimberley Process on several occasions, mainly by civil society representatives and the media, but the KP has been either slow to act, or has not acted at all. Smuggling of diamonds from Brazil, Venezuela, Guyana, and Zimbabwe has been debated at length, but have elicited weak, slow, or no response. The same has been true in cases where gross statistical anomalies suggest the need for urgent action: Guinea and Lebanon are two cases that were 'pending' throughout 2009, and which remain unresolved.[87]

'Technical assistance' has been used as a catch-all, last-minute answer to many of these problems. Assistance, regardless of how it is described, is not always the solution to problems of compliance. The KP approach, however, has been ad hoc and patchy. Guyana and Ghana, among others, are still awaiting technical assistance promised by the Kimberley Process. KP terminology and thinking need to expand beyond the idea of technical assistance as sending experts, to incorporate other ideas, including longer-term inputs and the provision of equipment.[88]

In sum, the Kimberley Process needs a rigorous, clear and phased compliance enforcement strategy that starts with assistance and internal pressure, moves to public naming and shaming, and then moves to high levels of sanctions, suspension, and expulsion.[89]

Smillie recommends the establishment of independent, third-party monitoring. He suggests that the KP Chair create a panel of experienced experts to design and propose a range of models for independent, third-party monitoring, complemented by rigorous

---

86   Ibid.
87   Ibid.
88   Ibid 11.
89   Ibid.

follow-up, credible sanctions in cases of continued non-compliance, and a decision-making process on non-compliance that is not hostage to political interference.[90]

Smillie also recommends the establishment of a small permanent KP secretariat to manage monitoring and follow-up, providing service to the KP Chair and working groups as required. The secretariat would not replace or supplant the Working Group on Monitoring; it would handle the organisational and managerial functions that currently fall to a single KP participant.[91]

Smillie also recommended the establishment of a multi-donor trust fund for timely and appropriate follow-up assistance in helping participants to meet KP minimum standards.[92]

Smillie also notes that the Kimberley Process has repeatedly ignored calls for the inclusion of oversight on the cutting and polishing industry in KP minimum standards. This sector remains vulnerable to, and a convenient laundry for, rough diamonds that have evaded KP scrutiny. The volume of illicit goods is growing: 100 per cent of Venezuela's production; conflict diamonds from Côte d'Ivoire; a large volume of Zimbabwe's diamonds moving through Mozambique; and an unknown volume of smuggled and stolen goods from other countries. Major seizures of illicit diamonds in India, Dubai, and elsewhere in recent months may be the tip of an iceberg.[93]

Smillie recommends that companies that cut and polish diamonds document their sources, and that their records be made subject to independent audit as an integral part of KP minimum standards. He suggests that the World Diamond Council should commission an independent evaluation of its system of warranties, to determine how it could improve the performance of industry actors in meeting KP challenges.[94]

Public transparency was originally a key focus of the Kimberley Process. The preamble to an early draft of the core document stated: 'Acknowledging that an international certification scheme for

---

90   Ibid.
91   Ibid.
92   Ibid.
93   Ibid 12.
94   Ibid 11.

rough diamonds will only be credible if supported by appropriate arrangements to ensure transparency and accountability with respect to its implementation.'[95]

Under the heading 'Cooperation and Transparency', however, the final KP core document lists seven provisions, dealing only with the exchange of information among participants. There is no discussion of public transparency. The KP's most notable failing in this area is the fact that reports of review visits are kept confidential.

The explanation given for this is that governments would not open themselves to full peer scrutiny if blemishes were to be made public. Most blemishes are, however, self-evident to inside observers, and are hardly a public secret. By hiding the reviews and their recommendations, and by failing to follow up on the recommendations, the KP effectively removes a tool that might improve matters without any effort on its part: publicity. Confidentiality, of course, also obscures the KP's lack of follow-up on its own recommendations. It also prevents concerned citizens from knowing about, and calling for change in, their governments' implementation of KP obligations.[96]

According to Smillie, greater openness in the Kimberley Process might be uncomfortable because it would be easier for the media, civil society, and others to hold it more accountable for timely follow-up on reviews, and for action on issues of serious non-compliance. But all of these stories find their way into the media anyway. Greater transparency would help to make the KP the regulatory body it aims to be, and the one the industry and African producer countries so badly require.[97]

Smillie recommends that all KP annual reports and reports of KP reviews, as a matter of course, be placed on the open KP website, along with details of follow-up action. A transparency working group should be established to develop criteria on exceptions to the rule, and to deal with special requests for confidentiality.[98]

---

95   Kimberley Process, (Working Document no 3/2001, 21 August 2001) cited in Smillie, I, 'Paddles for Kimberley: An Agenda for Reform' (Report, Partnership Africa Canada, June 2010) 12.
96   Smillie, I, 'Paddles for Kimberley: An Agenda for Reform' (Report, Partnership Africa Canada, June 2010) 12.
97   Ibid 14.
98   Ibid.

Smillie is also critical of what he considers self-censorship by the Kimberley Process of draft resolutions prepared for submission to the General Assembly of the United Nations. He states that, in 2009, Venezuela insisted that all references to Venezuela be dropped; China insisted that all references to human rights be dropped; and Zimbabwe insisted that all references to Zimbabwe be dropped. An anodyne UNGA resolution was passed, as a result, without a single reference to the issues that had most consumed the Kimberley Process over the previous two years. Echoing the growing dissidence from civil society and some industry players, several governments, including Switzerland, Sweden, Canada, and the United States, challenged the official KP version of events in the UNGA debate.[99]

According to Smillie, a major concern at the outset of KP negotiations was the potential cost implications of a global regulatory system. It was assumed that the industry would have to bear most of the cost, although as it turned out most of the financial burden has fallen on governments. In almost all countries, government has taken on most if not all of the cost of implementing the KP. The industry created the WDC to represent its interests in the Kimberley Process, and in some countries a low-cost chain of warranty system has been developed. Industry has participated in review visits and has contributed to the costs of special undertakings such as the 2006 review of Ghanaian diamond exports. All things considered, however, the cost of the KCPS to industry has been small.[100]

Civil society organisations have participated in all working groups, plenaries and intersessional meetings, and have participated in most review visits and missions. Civil society organisations have also undertaken a large number of independent reviews, studies and publications and have, arguably, borne a disproportionate cost of participation — and in holding the Kimberley Process accountable to its mandate.[101]

99   Ibid; Charbonneau, L, 'Zimbabwe "Blood Diamonds" Dispute Breaks out at UN', *Reuters Canada* (11 December 2009). Available at: ca.reuters.com/article/topNews/idCAT RE5BA3OI20091211.
100 Smillie, I, 'Paddles for Kimberley: An Agenda for Reform' (Report, Partnership Africa Canada, June 2010) 16–17.
101 Ibid.

The major cost implications lie in the adoption of an independent, third-party monitoring system, the establishment of a small secretariat to manage that function, and the required follow-up as an ongoing service to the chair of the day. The Working Group on Monitoring would continue to set the agenda and the policy framework and other working groups would remain unchanged. Smillie estimates the costs of the working group as being US$2.25 million per annum.[102]

Role of Industry

Although the primary regulator for the Kimberley Process at the national level is the national government and its agencies, one of the distinctive features of the Kimberley Process is that it can be considered as operating simultaneously as a government-regulated system as well as an exercise in industry self-regulation. The Kimberley Process has, from the outset, been driven by the needs and interests of the diamond industry. The De Beers corporation was a significant driving force in the finalisation of the Kimberley Process core document, and the industry as a whole has been represented through the WDC, an umbrella organisation for the large firm commercial diamond sector, at subsequent plenary meetings of the Kimberley Process.[103]

The journey of the high-end corporate diamond sector from the targets of bad press to advocates for the continued operation of the Kimberley Process is one of the striking features of the history of the organisation. One commentator has described this transformation as being a process of socialisation from self-interest to 'enlightened self-interest'.[104] The high sensitivity of the industry to its media and public image can be understood in the light of the end products of the industry. Although mining rough diamonds is undertaken for industrial purposes, such as their use in high-powered cutting tools, the bulk of the commercial value in the industry resides in the jewellery retail sector. Fundamentally, the value of a diamond in this context is aesthetic and sentimental, meaning that its value is at risk if it were

---

102 Ibid 17.
103 Commentator C Kantz argues that, while it was NGOs who highlighted the problem of conflict diamonds, the diamond industry became socialised to take responsibility for addressing the issue: Kantz, C, 'The Power of Socialization: Engaging the Diamond Industry in the Kimberley Process' (2007) 9(3) *Business and Politics* Art 2, 3.
104 Ibid.

to become associated with negative sentiment resulting from a link with human rights violations. Given its awareness of the risk, the commercial diamond industry has positioned itself as an advocate for the Kimberley Process, so as to separate its product from the conflict diamonds trade. One of the interesting findings from interviews conducted with Rio Tinto and the Australian Government was that Rio Tinto was a very strong advocate for the continued operation of the Kimberley Process whereas the Australian Government expressed the view that its relevance was diminished given the emergence of peace in Angola and Sierra Leone. The expression of this view by government would appear to be connected to the fact that much of the operating costs of the Kimberley Process comes from government rather than industry. This view, however, appears a little short-sighted given the emergence of new conflict diamonds threats in the DRC, Côte d'Ivoire, and Zimbabwe, in tandem with the importance of the Kimberley Process for the prevention of threats emerging again in Angola and Sierra Leone, or surfacing in another country.[105]

Although the WDC has observer status rather than voting status in the Kimberley Process, it can be argued that its representations at this level have a strong influence, whether voiced through the plenary or particular working groups and committees. This influence is particularly strong considering that decisions of the Kimberley Process plenary must be through consensus.[106] This means that, should the WDC lobby only a single government delegation to support its viewpoint, then it would not be possible for the Kimberley Process Plenary to make a decision regarding that which it does not concur.

The involvement of industry representatives and NGOs in the international Kimberley Process system is perhaps most pronounced through the committee and working group system. All committees have, both in principle and practice, involved industry and NGO representatives. Representatives are also mandated to participate in

105 Interview with Rio Tinto Representative (telephone interview, 30 May 2007); Written Response from Australian Government to Author's Interview Questions, 14 September 2007.
106 Kimberley Process, *The Kimberley Process Certification Scheme* (Core Document, 2002) s VI (5). Other international regulatory initiatives have sought to directly manage the activities of multinational corporations, including instruments such as the UN Norms of Responsibility of Transnational Corporations and Other Business Enterprises with Regard to Human Rights, discussed in Tripathi, S, 'International Regulation of Multinational Corporations' (2005) 33(1) *Oxford Development Studies* 117, 126.

review visits and review missions as part of the peer-review mechanism employed by the Kimberley Process. This level of involvement stands apart from standard expectations of NGOs and civil society, which typically are excluded from direct participation in the execution of treaty body decisions, and do not normally enjoy such a high level of influence in decision-making processes.[107]

One of the benefits of industry involvement is its ability to provide experts who enhance the system procedurally and practically. Rio Tinto has been asked for advice by other diamond-producing mines regarding Kimberley Process compliance, and has happily offered this advice in the knowledge that the company's reputation is affected if diamond-mining operations by others are not Kimberley-compliant.[108]

Industry Involvement Through Self-Regulation

Given the direct involvement of major diamond-producing corporations, such as De Beers, in the WDC, the formal diamond sector has been strongly integrated into the self-management of conflict diamonds prevention standards. As such, De Beers management, for example, has sought to ensure that its component corporate entities are compliant with certification requirements and its other obligations under the Kimberley Process.

One of the challenges, however, has been the involvement of the informal diamond sector in this process. It is very important to engage the informal sector, predominantly representing alluvial diamond miners, as this sector has been so directly implicated in producing diamonds for militia groups. As it does not appear that their interests are represented through the WDC, it will perhaps fall to national

---

107 Grant, A J and I Taylor, 'Global Governance and Conflict Diamonds: The Kimberley Process and the Quest for Clean Gems' (2004) 93(375) *The Round Table* 385, 385–387.
108 Interview between Author and Rio Tinto Representative.

governments to engage with the informal sector. It is noteworthy that guidelines for such engagement are currently being developed at the international level through the Kimberley Process to assist governments.[109] This movement culminated in the attendance of representatives of artisanal miners at the 2007 Kimberley Process Plenary in Brussels. The Kimberley Process decided to establish an Artisanal Mining Working Group to especially meet the needs of this central sector.[110]

The ability of the diamond industry to regulate itself in relation to the chain of warranties from import to retail was the subject of a 2004 study of the retail sector. Diamond retailers in the US and the UK were targeted by a survey, coordinated through the efforts of Global Witness and Amnesty International, to determine the effectiveness of diamond industry self-regulation at the retail end of the trading chain. In particular, reference was made to commitments made by the industry in January 2003 to implement a code of conduct to prevent buying or selling conflict diamonds; to implement a system of warranties requiring that all invoices for the sale of diamonds and jewellery containing diamonds contain a written guarantees that diamonds are conflict-free; to keep records of the warranty invoices given and received, and for this to be audited and reconciled on an annual basis by the company's own auditors; and to inform company employees about the industry's policies and government regulations to combat the trade in conflict diamonds.[111] Unfortunately, the study found that implementation of the system of warranties at the point of retail was inconsistent and not fully functioning. In particular, the study showed that retailers, where they were in fact implementing the voluntary self-regulatory measures, were not taking sufficient precautions to ensure that their suppliers were providing Kimberley-compliant diamonds. The study made a number of recommendations with diamond retailers in mind. It recommended that strict criteria

---

109 Kovanda, Karel, Kimberley Process Chair, 'Valedictory Remarks of Mr Karel Kovanda, Kimberley Process 2007 Chairman' (Intersessional Meeting, Brussels, Belgium, 8 November 2007).
110 Ibid.
111 Global Witness and Amnesty International, 'Déjà vu: Diamond Industry still Failing to Deliver on Promises: Summary of UK and US Results of UK and US Results of Global Witness and Amnesty International Survey' (Report, Global Witness, October 2004).

be applied in the selection of suppliers and that third-party auditing procedures be adopted to ensure that policies are working effectively; that retailers provide written assurances to consumers stating that the diamonds they purchase are conflict-free, so that the system of warranties covers the entire supply chain, from point of mine to point of sale to the consumer; and that retailers carry out education and training on conflict diamonds and the Kimberley Process, and require it as a condition of employment, so that salespeople are fully informed about policies and communicate this to consumers in a transparent manner.[112]

Both the formal and informal diamond sectors have some ability to enforce standards in a self-regulatory context. For example, internal disciplinary boards can be established to hear complaints about breaches of Kimberley Process standards by diamond industry employees.

Role of Non-Governmental Organisations

Non-governmental organisations such as Global Witness and Partnership Africa Canada were pivotal in bringing the problem of conflict diamonds to the attention of the international community, leading to the establishment of the Kimberley Process. They have subsequently had a vital role within the Kimberley Process, and have been described as representing the 'conscience of the Kimberley Process'.[113]

Non-governmental organisations play a very important role in the monitoring of Kimberley Process obligations both within the system and outside the system. Within the system, as discussed earlier, NGOs are able to bring potential non-compliance issues to the attention of the Kimberley Process, and to participate in the peer-review mechanism.

---

112 Global Witness, 'Broken Vows: Exposing the "Loupe" Holes in the Diamond Industry's Efforts to Prevent the Trade in Conflict Diamonds' (Report, March 2004).
113 Kovanda, Karel, 'Valedictory Remarks of Mr Karel Kovanda, Kimberley Process 2007 Chairman' (Intersessional Meeting, Brussels, Belgium, 8 November 2007).

NGOs have had a central role in the development of standard-setting within the Kimberley Process, including those standards relevant to regulation at the national level. They have an institutionalised role in relation to the functioning of the Kimberley Process, including providing ideas and recommendations as to further developments in the area of standard-setting.

The Kimberley Rules of Procedure clarify that observers may be invited to attend meetings of an ad hoc working group or a subsidiary body, either on a temporary or a permanent basis. The plenary may take a decision to revoke an invitation.[114]

One of the key successes and unique features of the Kimberley Process is its high degree of involvement with non-state entities, in particular representatives from the diamond industry and non-governmental organisations. NGOs play important roles in standard-setting and monitoring within the scheme. As stated in one review submission: 'NGOs and experts from throughout the diamond industry have played a vital role and their input is accepted (if not expected) as if they were states.'[115]

NGOs often initiate the incorporation of new standards in the context of Kimberley Process meetings. An example that stands out is the drive for more detailed internal controls to be set out through the Kimberley Process. In the area of monitoring, Global Witness and Partnership Africa Canada have contributed assessments as part of the three-year review of the Kimberley Process, they serve on the Working Groups on Monitoring and Participation within the Kimberley Process, and they have been involved in several review visits and missions.

---

114 Kimberley Process, *Rules of Procedure of Meetings of the Plenary, and its Ad Hoc Working Groups and Subsidiary Bodies* (2003).
115 Wexler, P, *The Kimberley Process Certification Scheme on the Occasion of its Third Anniversary: An Independent Commissioned Review* (Review Report submitted to the Ad Hoc Working Group on the Review of the Kimberley Process, February 2006) 6.

Rather than preferring an adversarial approach to the Kimberley Process, these organisations engage dynamically within the process to effect change.[116]

In their landmark study of the 'governance triangle', assessing a range of tripartite (government, business, NGO) initiatives, Abbott and Snidal interestingly position the Kimberley Process at the centre of their triangle model, indicating their assessment that, as regards the relative influence of the three stakeholders, the KP is perhaps the most evenly balanced of the assessed initiatives. It is possible to challenge this assessment, however, by observing that only governments are voting members of the KP, despite the significant influence of NGOs and businesses. Given the walk-out at the Kinshasa interplenary meeting, and the subsequent non-attendance by NGOs at the November 2011 plenary, the structural weighting towards governments has more clearly made itself manifest.[117]

---

116 Kimberley Process, *The Kimberley Process Certification Scheme* (Core Document, 2002) ss III, IV. Work has been initiated within the Kimberley Process in relation to the effectiveness of such internal controls and industry self-regulation. Kimberley Process, 'Final Communique: Kimberley Process Plenary Meeting' (Gatineau, Canada, 29 October 2004). This culminated with Kimberley Process Plenary's endorsement of a document, called the Brussels Declaration, on internal controls of participants with rough diamond trading and manufacturing. The declaration gives guidance on controls for record keeping, spot checks of trading companies, physical inspections of imports and exports, and maintenance of verifiable records of rough diamond inventories. Kimberley Process, '2007 Kimberley Process Communique' (Brussels, Belgium, 8 November 2007). A critique of the lack of internal controls in the Kimberley Process is set out in Fishman, J L, 'Is Diamond Smuggling Forever?: The Kimberley Process Certification Scheme: The First Step Down the Long Road to Solving the Blood Diamond Trade Problem' (2005) 13 *University of Miami Business Law Review* 217, 237–238; Interview between Author and Global Witness Representative. The use of strategies within and outside formal processes to influence outcomes by NGOs has been described as 'multitrack diplomacy'. Grant, A J and I Taylor, 'Global Governance and Conflict Diamonds: The Kimberley Process and the Quest for Clean Gems' (2004) 93(375) *The Round Table* 385, 386–687.
117 Abbott, K and D Snidal 'The Governance Triangle' in W Mattli and N N Woods (eds), *The Politics of Global Regulation* (Princeton University Press, 2009) 50–52; Partnership Africa Canada, 'Kimberley Process lets Zimbabwe off the Hook Again', (2 November 2011). Available at: www.pacweb.org/Documents/Press_releases/2011/KP_lets_Zim_off_the_hook_Nov2011.pdf.

Role of NGOs External to the Kimberley Process

Outside the system, NGOs have also been effective in undertaking their own independent monitoring of the effectiveness of the Kimberley Process in particular contexts and countries.[118] It was through reports by organisations such as Global Witness and Partnership Africa Canada that world attention was given to the conflict diamonds problem, and Global Witness has continued to provide this external aspect of scrutiny of the Kimberley Process system. NGOs have proven adept at attracting international media attention to the conflict diamonds issue. The 1998 protests organised by Global Witness in front of Tiffany's jewellery store in New York was a landmark in attracting global attention to the problem.[119] Another media event occurred in 2006 with the release of the popular Hollywood movie *Blood Diamond*, which brought further public attention to the issue.[120] Considering this external aspect of its scrutiny, recent reports by the organisation have considered the ability of the diamond industry to implement its chain of warranties from the point of import to the point of retail. A report dated June 2010 released by Partnership Africa Canada provides an important external and contemporary critique of the operation of the Kimberley Process.[121]

Bringing conflict diamonds trading to the attention of the global media, and naming and shaming unscrupulous corporate or individual behaviour, acts itself as a form of enforcement action, which is readily available to non-governmental organisations. The term 'enforcement' is used here in the regulatory sense, rather than in the normal legal sense that generally connotes action by a central authority.[122] Enforcement in this sense indicates the pressure that negative media attention brings to bear on industry, which influences industry to change its behaviour to conform with international regulatory standards. It is

---

118 Examples of public reports are Global Witness, 'Broken Vows: Exposing the "Loupe Holes" in the Diamond Industry's Efforts to Prevent the Trade in Conflict Diamonds' (Report, March 2004); and Partnership Africa Canada, 'Diamonds and Human Security: Annual Review 2008' (Annual Report, October 2008).

119 Bieri, F, *From Blood Diamonds to the Kimberley Process: How NGOs Cleaned Up the Diamond Industry* (Ashgate Publishing Ltd, 2010) 41.

120 Ibid 148.

121 Smillie, I, 'Paddles for Kimberley: An Agenda for Reform' (Report, Partnership Africa Canada, June 2010).

122 As discussed in Chapter 5, regulation is theorised to consist of the processes of standard-setting, monitoring, and enforcement.

noted that the ultimate consumers of diamond products purchase diamond jewellery as much for its image as its intrinsic qualities. Should diamonds develop a bad media image, the value of the industry would rapidly decrease. NGOs may also use the threat of poor media publicity as leverage with industry and governmental groups to seek higher standards under the Kimberley Process.[123]

The Diamond Development Initiative International

The quality of internal control systems varies greatly from one KP participant to another. Implementing effective internal controls is most difficult for countries with alluvial diamond reserves. Alluvial diamonds are found in vast areas, usually along riverbeds, and these areas are often beyond the control of both states and the diamond industry. Alluvial diamonds are most frequently mined by artisanal miners who work with simple tools and often earn less than US$1 per day. Kimberlite diamonds, on the other hand, are mined with capital-intensive machinery that extracts the diamonds directly from a volcanic pipe. Botswana, the exemplar for positive African development based on diamonds, extracts its gems from kimberlite mines. African countries with large alluvial diamond reserves include Angola, CAR, Côte d'Ivoire, the DRC, Guinea, Ghana, Liberia, and Sierra Leone.[124]

Artisans mine between 10 and 20 per cent of the diamonds used for jewellery, which makes them an important part of the industry. They have no better employment opportunities and typically hope to make a big find in the diamond fields. But the vast majority, despite their back-breaking labour, will never find that large stone. The diamonds they dream of are referred to as poverty diamonds, but are closely linked to conflict diamonds: 'The poverty, the hundreds of thousands of willingly exploited adults and children, and the

---

123 Tamm, I J, 'Dangerous Appetites: Human Rights Activism and Conflict Commodities' (2004) 26 *Human Rights Quarterly* 687, 691. Academic commentators have noted, however, that negative publicity can operate indiscriminately, thereby damaging the diamond revenues of producer nations that are considered to have 'clean' diamond industries, such as Botswana. Taylor, I and G Mkhawa, 'Not Forever: Botswana, Conflict Diamonds and the Bushmen' (2003) 102 *African Affairs* 261, 271. See also Shaw, Timothy M, 'Regional Dimensions of Conflict and Peace-Building in Contemporary Africa' (2003) 15 *Journal of International Development* 487, 492.

124 Bieri, F, *From Blood Diamonds to the Kimberley Process: How NGOs Cleaned Up the Diamond Industry* (Ashgate Publishing Ltd, 2010) 149.

volatility of the diamond fields make for a highly flammable social cocktail, one that has ignited several times in recent years, with tragic results.'[125] The alluvial sector is closely linked to civil wars in those countries and has led to regional and international instability. Rebels continue to control alluvial diamond fields and poor artisanal diamond workers are easily recruited for rebel armies or to sell the diamonds they find to regional warlords. A range of social problems is associated with artisanal mining: child labour, HIV infections, environmental destruction, crime and violence, poverty, and unhealthy and dangerous working conditions. Some states have declared the alluvial diamond mining sector illegal, while most states neglect it altogether, thereby exacerbating the problems. 'There are no cases in Africa where artisanal diamond mining has been supported and regulated successfully.'[126]

The conflict diamonds campaign, in particular that by Global Witness and Partnership Africa Canada, did not pay close attention to development issues relating to alluvial miners until late 2004. But local African initiatives appeared as early as the beginning of 2002, as soon as relative stability emerged in the region. Various pilot projects attempted to address the issue. For example, the government of Sierra Leone created the Diamond Area Community Development Fund, through which part of the export tax on diamonds is directly returned to the artisanal diamond communities. The DRC created an organisation to assist in creating mining cooperatives. The Campaign for Just Mining was launched by the Network Movement for Justice and Development (a Sierra Leonean NGO and Southern partner organisation of Partnership Africa Canada) in January 2000, following the widely publicised Partnership Africa Canada report *Heart of the Matter* on Sierra Leone. The Campaign for Just Mining's goals are to 'promote sustainable development in Sierra Leone by advocating accountability, transparency and social responsibility within the

---

125 Global Witness and Partnership Africa Canada, 'Rich Man, Poor Man, Development Diamonds and Poverty Diamonds: The Potential for Change in the Artisanal Alluvial Diamond Fields of Africa' (Report, October 2004) 6–7.
126 Global Witness and Partnership Africa Canada, 'Rich Man, Poor Man, Development Diamonds and Poverty Diamonds: The Potential for Change in the Artisanal Alluvial Diamond Fields of Africa' (Report, October 2004) 8; Bieri, F, *From Blood Diamonds to the Kimberley Process: How NGOs Cleaned Up the Diamond Industry* (Ashgate Publishing Ltd, 2010) 149–150.

mining sector'.[127] The campaign is squarely rooted in a human rights frame, or, as they call it, a 'rights-based approach to mining', including the right to a sustainable livelihood (i.e. formalising diggers' employment status); the right to basic services; the right to security; protection of the environment; ensuring long-term human security, including food security; and avoiding illnesses such as malaria; and the right to participate in various decision-making processes.[128]

Many of these local initiatives were funded through bilateral aid, which was often forthcoming after peace agreements had been reached. The Peace Diamond Alliance was launched in December 2002, organising diggers in cooperatives to help them obtain better prices for diamonds. It was funded by the United States Agency for International Development (USAID) and managed by Management Systems International, a Washington-based consulting firm. 'It has brought together an eclectic group of local and international NGOs, diamond buyers, mining companies and government officials.'[129] Other pilot projects were also initiated by state donor agencies and NGOs. Bilateral agencies such as USAID and some international governmental organisations, notably the World Bank, committed funds and organisational capacity to the issue of artisanal and small-scale mining. Many of these initiatives grew out of foreign policy efforts by states, especially the USA and the UK, sometimes in relation to their brokering of peace agreements. For instance, the involvement of the US in Sierra Leone in 1999 brought USAID and other bilaterally focused agencies into the region. The civil wars in the area, particularly in Sierra Leone, had become closely associated with diamonds during the course of the conflict diamonds campaign. The mining initiatives emerged in the context of bilateral agencies' attempts to address one of the primary causes of the wars.[130]

---

127 Global Witness and Partnership Africa Canada, 'Rich Man, Poor Man, Development Diamonds and Poverty Diamonds: The Potential for Change in the Artisanal Alluvial Diamond Fields of Africa' (Report, October 2004) 11.

128 Global Witness and Partnership Africa Canada, 'Rich Man, Poor Man, Development Diamonds and Poverty Diamonds: The Potential for Change in the Artisanal Alluvial Diamond Fields of Africa' (Report, October 2004) 12; Bieri, F, *From Blood Diamonds to the Kimberley Process: How NGOs Cleaned Up the Diamond Industry* (Ashgate Publishing Ltd, 2010) 149–150.

129 Global Witness and Partnership Africa Canada, 'Rich Man, Poor Man, Development Diamonds and Poverty Diamonds: The Potential for Change in the Artisanal Alluvial Diamond Fields of Africa' (Report, October 2004) 11.

130 Bieri, F, *From Blood Diamonds to the Kimberley Process: How NGOs Cleaned Up the Diamond Industry* (Ashgate Publishing Ltd, 2010) 150–151.

These initiatives reveal several important characteristics. First, they were frequently organised in a tripartite fashion. Multi-stakeholder models had been firmly established in the development/aid sector by the late 1990s, meaning that national and local civil society partners were usually involved. In the case of mining extraction activity, where state and large-scale industry are frequently co-owners, it meant that large-scale industry was also an important stakeholder for small-scale mining initiatives.[131]

Second, the bilateral funds that were disbursed for artisanal diamond mining projects showed that the issue of poverty or development diamonds was already on the agenda of some state donor agencies by 2002–03. For example, the World Bank's Community and Small Scale Mining initiative, launched in March 2001, had been in the works since September 1999. It focused largely on artisanal mining in South America and Asia. But government donor agencies' agendas were also affected by the conflict diamonds campaign in early 2000. For example, the third annual general meeting of the World Bank's Community and Small Scale Mining Project was held in Ghana in September 2003, with diamond mining prominent on the agenda. Partnership Africa Canada gave a presentation on the KP. The aid that followed the peace agreements was intended to address the circumstances causing conflicts. The conflict diamonds campaign had made diamonds appear to be especially important as a source of conflict. The Sierra Leone war, for instance, had come to be defined almost entirely as a conflict diamonds issue. The conflict diamonds campaign thus influenced aid and development responses in 2002 and helped put artisanal diamond mining on the agenda of some governmental aid agencies.[132]

Third, the emergence of the development diamond projects shows that the development frame was on the agenda of local NGOs in former conflict diamonds areas. For many African NGOs, in fact, these developmental aspects relating to artisanal miners were always at the forefront, and many African NGOs were eager to shift the focus of the conflict diamonds campaign towards underlying issues of development. Prior to 2004, African NGOs were unable to add the development framework to the global public awareness campaign, and the Western NGOs leading these efforts focused narrowly on conflict

---

131 Ibid 151.
132 Ibid.

diamonds. However, African NGOs became partners of bilateral aid agencies eager to work with civil society on local development projects on small-scale mining. That is how they were able to pursue a broader, more holistic approach to the role of diamonds in conflict regions and to get the Western NGOs to broaden their focus.[133]

The first strategic meeting leading to the establishment of the Diamond Development Initiative International was called by Global Witness, Partnership Africa Canada, and De Beers in January 2005 in London. This meeting was chaired by former US Assistant Secretary of State for Africa Walter Kansteiner, and included representatives from states, the European Community, the United Nations, the Department for International Development (British Foreign Aid Agency), USAID, the World Bank, industry, and NGOs.[134]

A second meeting took place in Washington DC in June, in conjunction with a Communities and Small Scale Mining meeting at the World Bank. Both meetings were devoted to defining the goals and scope of the initiative. The DDII was to address the political, social, and economic challenges associated with artisanal diamond mining in Africa, attempting to bring this large informal sector into the formalised economy. While some industry players remained sceptical, De Beers and Martin Rapaport had already been involved in the Peace Diamond Initiative in Sierra Leone. They had been important in bringing the conflict diamonds issue to the industry's attention in early 2000, when the campaign got underway. De Beers' motives for its proactive involvement on development diamonds were viewed with suspicion by some NGOs and by some in the industry. However, other NGOs recognised that, while De Beers was operating from the perspective of the financial interests of their company by managing potential bad publicity, this could be seen as a form of 'enlightened self-interest'.[135]

Initiators of the DDII (De Beers, Partnership Africa Canada, Global Witness, the Rapaport Group, and Jeffrey Davidson, representing the World Bank's Communities and Small Scale Mining Secretariat) were joined by the Foundation for Environmental Sustainability and

---

133 Ibid.
134 Ibid.
135 Ibid.

Security, and the International Diamond Manufacturer's Association, and 'the DDII [was] endorsed by the governments of Sierra Leone, Guinea, the DRC, Namibia and others, and ... received start-up project funding from Canada's Department of Foreign Affairs'.[136] Overall, alluvial diamond mining countries expressed great interest in the initiative, while other diamond mining countries, such as South Africa, Botswana, and Namibia, were less engaged in the matter.[137]

Communities and Small Scale Mining's experience in dealing with artisanal mining suggested that getting donors involved in such initiatives was difficult. It was explained that extractive industries were regarded by some donors as an area too complicated to get into. Mining was not liked generally, and artisanal mining was traditionally considered particularly problematic by donors.[138]

Despite some of these early challenges, participation in the inaugural DDII meeting held in Accra (Ghana) on 27–30 October was very good. It was limited to 80 representatives from states, industry, and civil society. The meeting was financed by registration fees from industry and northern states ($400 each), $30,000 from the World Bank, and $5,000 each from Rapaport, De Beers, Global Witness, and Partnership Africa Canada. Registration for NGOs and Southern states was free. At the Accra meeting, the DDII's goals were further developed.[139]

The DDII seeks to integrate artisanal diamond mining initiatives already underway, such as the above-mentioned projects — the Peace Diamond Alliance, Communities and Small Scale Mining, Campaign for Just Mining. It shares with the Kimberley Process the vision

---

136 Partnership Africa Canada, *Diamond Development Initiative International*. Available at: www.pacweb.org.
137 Gizenga, Dorothee Ngolo, interview with F Bieri, 6 July 2006 in Bieri, F, *From Blood Diamonds to the Kimberley Process: How NGOs Cleaned Up the Diamond Industry* (Ashgate Publishing Ltd, 2010) 155.
138 Smillie, Ian, interview with F Bieri, 6 July 2006, in Bieri, Franziska, *From Blood Diamonds to the Kimberley Process: How NGOs Cleaned Up the Diamond Industry* (Ashgate Publishing Ltd, 2010) 155.
139 Smillie and Ngolo Gizenga, interview with F Bieri, 6 July 2006, in Bieri, Franziska, *From Blood Diamonds to the Kimberley Process: How NGOs Cleaned Up the Diamond Industry* (Ashgate Publishing Ltd, 2010) 155–156.

of being a multi-stakeholder partnership between governments, NGOs, and industry, seeking to pool resources, experience, and knowledge, although without the formal organisational structure of the KP.[140]

The DDII seeks not only to put artisanal diamond mining on the agenda of NGOs and donors working on mining issues (such as the World Bank's Communities and Small Scale Mining project) but also to create an encompassing approach that is not country — or initiative — specific. It attempts to translate these initiatives into a more cohesive and global frame in which diamonds are a development issue, and to get the big international development organisations on board.[141]

In essence, the goal is making the world's 13 million artisanal diamond miners, most of whom live in total poverty, and face multiple health, peace, and security challenges, a new focus in development. Artisanal diamond miners tie the conflict and development frames together. By consolidating local initiatives, the DDII forms a bridge between the conflict frame and the development frame.[142]

Initially, Partnership Africa Canada and Global Witness did not explicitly assign any staff, not even part-time, to the initiative.[143] The lack of resources is partially explained by the fact that the KP continues to exhaust NGOs' capacities. It is more difficult to engage other NGOs and donors on the development diamonds than conflict diamonds.[144]

Ngolo Gizenga from Partnership Africa Canada mentions several hindering factors. First, it is more difficult to garner support from other NGOs because development diamonds do not lend themselves to the same 'sexy' consumer campaign as conflict diamonds. Second, there was suspicion about working on an initiative jointly with the industry and states. Specifically, Partnership Africa Canada was

---

140 Diamond Development Initiative International, *Accra Conference Background Note*, (October 2005) 2. Available at: www.artisanalmining.org/userfiles/file/DDI_Accra_Oct.5.pdf.

141 Smillie, quoted in Partnership Africa Canada, *Other Facets*, Ottawa (October 2006) 1.

142 Bieri, F, *From Blood Diamonds to the Kimberley Process: How NGOs Cleaned Up the Diamond Industry* (Ashgate Publishing Ltd, 2010) 156.

143 Smillie, Ian, interview 6 July 2006 with F Bieri, in Bieri, Franziska, *From Blood Diamonds to the Kimberley Process: How NGOs Cleaned Up the Diamond Industry* (Ashgate Publishing Ltd, 2010) 156.

144 Ibid 156; Gizenga, Dorothee Ngolo, interview with F Bieri, 6 July 2006 in Bieri, F, *From Blood Diamonds to the Kimberley Process: How NGOs Cleaned Up the Diamond Industry* (Ashgate Publishing Ltd, 2010) 157.

accused of getting into bed with De Beers. However, building on the track record of cooperation from the Kimberley Process experience, which was also tripartite, the DDII seemingly overcame this obstacle and was able to continue to attract financial support for its activities. The personal relationships developed in the KP served in good stead for the creation of the DDII. The same individuals involved in the DDII have sat through countless hours of negotiations in the KP and have socialised informally at frequent meetings around the world.[145]

While particular industry players are involved in the DDII, most notably De Beers and the Rapaport group, the WDC — the industry NGO created to deal with the conflict diamonds issue — remained uninvolved. While formally welcoming the initiative, the WDC kept its distance, saying that it was created to address only the conflict diamonds issue.[146]

The WDC thus was intent on keeping its focus on the KP. Its distance from the DDII can possibly be explained by the fact that the DDII was an initiative of De Beers, which is both loved and hated for its dominant position in the industry. Also, the DDII leaves out industry members that have little to do with artisanal mining (i.e. industries involved in kimberlite extraction, many traders and retailers) or who were already reluctant partners in the KP/WDC.[147]

Despite close connection between the DDII and the KP, there was never any intention of incorporating the DDII within the KP. States, the industry, and the NGOs agreed that this would have been counter-productive. No one wanted to jeopardise what had so far been achieved in the KP by overburdening it. In addition, the KP would have prevented the DDII from engaging state donor agencies who could help fund the initiative. As explained in the chapter on implementation, the KP is minimally funded, and state officials

---

145 Gizenga, Dorothee Ngolo, interview with F Bieri, 6 July 2006 in Bieri, F, *From Blood Diamonds to the Kimberley Process: How NGOs Cleaned Up the Diamond Industry* (Ashgate Publishing Ltd, 2010) 157; Bone, Andrew, interview with F Bieri, 7 July 2005 in Bieri, F, *From Blood Diamonds to the Kimberley Process: How NGOs Cleaned Up the Diamond Industry* (Ashgate Publishing Ltd, 2010) 160; Bieri, F, *From Blood Diamonds to the Kimberley Process: How NGOs Cleaned Up the Diamond Industry* (Ashgate Publishing Ltd, 2010).

146 Izhakoff, interview with F Bieri, 2005 in Bieri, F, *From Blood Diamonds to the Kimberley Process: How NGOs Cleaned Up the Diamond Industry* (Ashgate Publishing Ltd, 2010) 160.

147 Bieri, F, *From Blood Diamonds to the Kimberley Process: How NGOs Cleaned Up the Diamond Industry* (Ashgate Publishing Ltd, 2010) 160–161.

in the KP are not connected to the state agencies most likely to be potential donors; instead, they represent trade departments or mineral extraction ministries. Third, most KP participants are not involved in the DDII. While any diamond trading state must participate in the KP to engage in international diamond transactions, most states take no interest in and are sceptical about being asked to engage on issues of diamonds and development.[148]

NGOs identified that the KP was a regulatory system but not a tool for development, which was the emerging role of the DDII.[149] The DDII reminded KP participants why the certification was initially launched. Important links between the KP and the DDII existed and were nurtured. The KP provided the networks, reputations and know-how, especially for tripartite interaction, which were applied in the DDII.[150]

The KP also lent the DDII legitimacy by commending its activities, which is itself a sign of the global esteem the KP had achieved by 2005. On several occasions, the DDII was given the opportunity to make presentations at formal KP meetings; Ian Smillie presented the DDII to the Moscow plenary in November 2005. The final communiqué of the KP Moscow plenary concluded: 'Liaison between the KP and the DDI was encouraged in order to optimize synergies.'[151] The DDII thus kept development issues on the agenda of the KP by closely linking development to conflict diamonds and showing that the conflict diamonds problem could not be solved without also addressing development issues related to artisanal miners.[152]

Most importantly, the DDII addressed one of the KP's key weaknesses: the lack of internal controls in alluvial diamond states. The KP in essence doesn't capture much alluvial diamond mining. It is unable to assess with confidence from which mine or mining area an alluvial

---

148 Ibid 161.

149 Partnership Africa Canada, *Diamond Development Initiative International*. Available at: www.pacweb.org.

150 Bieri, F, *From Blood Diamonds to the Kimberley Process: How NGOs Cleaned Up the Diamond Industry* (Ashgate Publishing Ltd, 2010) 161.

151 Kimberley Process, *Kimberley Process Plenary Meeting Final Communique*, (Moscow, 2005). Available at: www.kimberleyprocess.com/en/2005-final-communique-plenary-moscow compendium.

152 Bieri, F, *From Blood Diamonds to the Kimberley Process: How NGOs Cleaned Up the Diamond Industry* (Ashgate Publishing Ltd, 2010) 161.

diamond comes. Concerned about this weakness, the KP established an ad hoc working group on artisanal mining, estimating that the concerns of an estimated 1 million artisanal miners were not being addressed adequately.[153]

Thus, for the KP, the DDII is an important means of tackling an issue that threatens the effectiveness and legitimacy of the KP. In broader terms, the DDII is important because the KP alone cannot ensure peace in the region. The KP operates in a context supported by UN Peacekeeping operations in Liberia, Côte d'Ivoire, and the Congo, with a combined troop strength of 38,000 and an annual budget of US$2.3 billion. The DDII contributes a further layer to the peacekeeping structure.[154]

Current arrangements are unsustainable in the long run, and without the UN peace forces in the Congo the Kimberley Process cannot effectively ensure that diamonds will not fuel renewed conflicts. This may explain why the DDII received much attention and was positively endorsed by the KP. That the DDII's goals were defined as complementary and supportive of the KP facilitated a sound relationship between the two. The DDII did not undermine the KP for its ineffectiveness with regard to development, which would have delegitimised the process as a whole.[155]

Overall, the KP served as an important starting point for the DDII, but the DDII was built as a separate organisational effort. While relationships developed in the KP were important, not all industry players involved in the KP became involved in the DDII. For instance, the WDC remained focused on the KP, while De Beers and Rapaport helped initiate the DDII. States that were crucial in setting up the KP (South Africa, Botswana, Namibia) were not closely involved in the DDII, while alluvial diamond-producing nations and several donor countries became active in the DDII. Canada, Britain, and the United States were involved in both, though for Britain and the United States different state agencies dealt with the KP and development diamond issues (the US Trade Department and the European Community in the KP, USAID and DFID in DDII), while for Canada the Department

---

153 Diamond Development Initiative International, *Accra Conference Background Note,* (October 2005) 5. Available at: www.artisanalmining.org/userfiles/file/DDI_Accra_Oct.5.pdf.
154 Smillie, quoted in Partnership Africa Canada, *Other Facets,* Ottawa (October 2006) 1.
155 Ibid 162.

of Foreign Affairs was the central agency for both initiatives. In 2008, funding for the DDII was provided by the government of Sweden, Tiffany & Co. Foundation, Partnership Africa Canada and the JCK Industry Fund, with an annual budget of $287,580.[156]

In the end, the NGOs did not press for the DDII agenda to become incorporated into the operations of the KP. Whilst some view this separation as desirable, it is arguable that the KP would be enhanced and revitalised by adopting a broader mandate, as exemplified by the DDII. This matter is discussed in more depth below. See in particular the discussion in Chapter 7 and Chapter 8.

DDII's first operational year, 2008, saw important progress made. The DDII produced several 'Standards and Guidelines' materials, offering various stakeholders important information on artisanal mining in specific countries. The DDII engaged in a pilot study on Guyana's registration system of alluvial miners and its internal diamond production tracking mechanisms. This study now serves workshops and training sessions in Africa to implement similar systems there.[157]

Concluding Remarks

This chapter has provided an overview of the Kimberley Process, including the main ways in which it operates. It discussed procedures for membership, annual reporting, and the peer-review mechanism, including the manner in which the Kimberley Process has dealt with cases of serious non-compliance. The chapter has also considered the role of industry in the Kimberley Process, including self-regulation, as well as the role of NGOs, including those NGOs external to the KP, with a particular focus on the DDII. The chapter has also provided an evaluation of the overall effectiveness of the KP. The KP has been successful in creating a unique regulatory organisation harnessing the unique contributions of governments (legitimacy, capacity for national enforcement), industry (technical knowledge and self-regulation), and

156 Diamond Development Initiative International, *2008 Annual Report: Beyond Dreams the Journey Begins* (2008) 9; Bieri, F, *From Blood Diamonds to the Kimberley Process: How NGOs Cleaned Up the Diamond Industry* (Ashgate Publishing Ltd, 2010) 162–163.
157 Diamond Development Initiative International, *2008 Annual Report: Beyond Dreams the Journey Begins* (2008) 9; Bieri, F, *From Blood Diamonds to the Kimberley Process: How NGOs Cleaned Up the Diamond Industry* (Ashgate Publishing Ltd, 2010) 162–163.

NGOs (widespread, objective networks for monitoring and behaviour modification). Its success in enlisting the support of major industry groups, in particular cartel leader De Beers, is particularly notable. On the issue, De Beers has done a 180-degree turn from opposition to ferocious agreement with NGO players, perhaps aware of focusing attention away from the potential criminal liability of its officers for complicity in human rights abuses that occurred in Angola and the DRC during the 1990s. With De Beers and peak group WDC taking ownership of the issue, a large piece of the solution falls into place simply by having these major players not engaged in the purchasing of conflict diamonds.

The KP has created unprecedented transparency in the diamond industry, particularly in the provision of statistics, which are more comprehensive, reliable, and accessible than before. With reference to these statistics, it is possible to estimate that conflict diamonds have fallen to less than 1 per cent of the international diamond trade. While the reduction in conflict diamonds is largely attributable to the emergence of peace in countries such as Sierra Leone and Liberia, the KP has arguably contributed to the tackling of the conflict diamonds problem and creating these conditions of peace.

Despite the level of success it has achieved, the KP has struggled recently in the face of serious non-compliance by some of its government members. Despite acting decisively to protect the integrity of the KP when faced by large-scale illicit trade being funnelled through the RCB in 2004, the KP did not act in this way when Venezuela behaved in a similar fashion in 2006. Most disturbing, however, are the large-scale killings and rapes that have occurred in Angolan and Zimbabwean artisanal fields, without eliciting suspension or expulsion from the KP. Diamonds emerging from these environments must be classified as conflict diamonds, and the continuing failure to act by the KP is a potential threat to its long-term viability.

4

Kimberley at the National Level: Fancy Footwork?

The people back home wouldn't buy a ring if they knew it cost someone else their hand.

From the movie *Blood Diamond*[1]

Chapter Overview

National governments, particularly those of diamond-producing nations, are the front-line defences against conflict diamonds. At the heart of the KP is the initial certification by the country of first export that the diamonds are, in fact, free from association with international human rights crimes. For this certification to be meaningful, an effective system of internal controls from the diamond mining site to customs at the place of international export is required. This chapter focuses on those internal controls as well as, to some extent, the internal controls required by cutting and polishing nations, and those nations that are primarily involved in the retail of diamond jewellery. In doing so, the chapter makes an important contribution towards answering the first research question, namely, to what extent has the conflict diamonds governance system achieved its objectives? The chapter considers in detail the manner in which several African

---

1   *Blood Diamond*, 2006, motion picture, Warner Bros, www.imdb.com/title/tt0450259/quotes.

diamond-producing nations — Angola, the Democratic Republic of Congo (DRC), Côte d'Ivoire, and Sierra Leone — have incorporated internal controls into domestic legislation. By way of contrast with a developed country, the implementation of KP obligations by Australia is also discussed. Finally, legal developments in the Netherlands are considered, most notably the efforts of that government to hold accountable one of its nationals (Gus Kouwenhoven) before its national court system, in relation to resource-based crimes under international law. The chapter closes following concluding remarks.

Implementation at the National Level

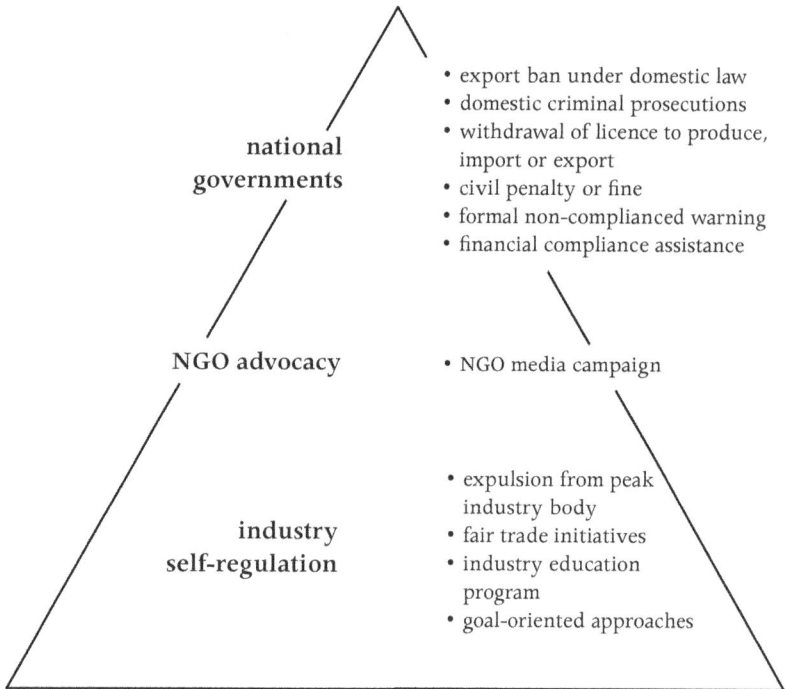

Figure 4.1: Kimberley Process: The Domestic Regulatory Pyramid
Source: Author's research.

The Kimberley Process is premised on the idea of a chain of warranties that verifies the origin of rough diamonds from the point of mining to the point at which the diamonds are purchased by consumers at retail outlets. The primary obligations under the Kimberley Process relate to the regulation of the import and export of conflict diamonds. Each country participant, on export, is required to ensure that each shipment of rough diamonds is accompanied by a duly validated certificate. On import, each participant must require a duly validated certificate be presented by the importer. The participant must also send confirmation of receipt to the exporting authority.[2] Governments must ensure that no shipment of rough diamonds is imported or exported without such certification, thereby excluding non-participating countries from the rough diamond trade. Transit countries, however, are not required to verify or issue certificates, providing that parcels are not tampered with in transit.

Even where countries have implemented these obligations into their domestic laws,[3] such regulation will only apply to the process of exporting or importing diamonds. The Kimberley Process provides that regulation of the initial half of the trade in diamonds, from the point of mining to the point of export, must be guaranteed by states themselves, though little detail of how such internal controls are to be implemented is provided in the international instrument itself. Consequently, this matter has been left to individual states to devise, with varying degrees of success. Similarly, states are responsible for ensuring that rough diamonds that are imported are Kimberley-compliant.[4]

---

2   Kimberley Process, *The Kimberley Process Certification Scheme* (Core Document, 2002) s III.
3   The Australian Government made the point that the Kimberley Process border control measures must be implemented under the domestic laws or regulations of each participating country. Written Response from Australian Government to Author's Interview Questions (14 September 2007).
4   It is noted that work was initiated within the Kimberley Process regarding the effectiveness of internal controls and industry self-regulation. Kimberley Process, 'Final Communique: Kimberley Process Plenary Meeting' (Gatineau, Canada, 29 October 2004). This culminated with the Kimberley Process Plenary's endorsement of a document called the Brussels Declaration on internal controls of participants with rough diamond trading and manufacturing. The declaration gives guidance on requirements for record keeping, spot checks of trading companies, physical inspections of imports and exports, and maintenance of verifiable records of rough diamond inventories. Kimberley Process, '2007 Kimberley Process Communique' (Brussels, Belgium, 8 November 2007). A critique of the lack of internal controls in the Kimberley Process is set out in Fishman, J L, 'Is Diamond Smuggling Forever?: The Kimberley Process Certification Scheme: The First Step Down the Long Road to Solving the Blood Diamond Trade Problem' (2005) 13 *University of Miami Business Law Review* 217, 237–238. See also, Global Witness and Partnership Africa Canada, 'The Key to the Kimberley: Internal Diamond Controls: Seven Case Studies' (Report, 2004).

Spearheaded by NGOs, there has been significant pressure for the Kimberley Process Plenary to finalise an administrative decision that sets down particular requirements for internal controls in rough diamond–producing nations. For example, there are currently no specific requirements regarding the licensing of diamond mining and trading activities, or requirements that customs officials have the power to monitor the industry through spot-checks, or even requirements that fines and criminal penalties attach to trafficking black market diamonds. There should be minimum training standards for diamond valuers and customs officials, and requirements regarding how they report back about diamond exports. In the absence of such specifications, it is difficult to ensure that effective internal controls are in place. The task of review teams sent through the peer-review system is made more difficult as there are no established requirements to look for in assessing the internal controls of countries being investigated.[5]

For example, government monitoring officers are not able to monitor effectively on the ground at mining centres. The overall governance of the system is hindered by lack of implementation capacity, and corruption.

One dimension of the problem is that a lot of diamond miners are working in dangerous conditions for very little remuneration, which is an example of the many labour, social, and environmental issues underpinning the conflict diamonds problem. Many of these diggers are unregistered and unlicensed by the national government.[6]

Similar comments, but with a different emphasis, on the issue of internal controls were made by the representative from Rio Tinto in his interview response. The representative stated that regulation by national governments probably varies quite a lot and can be improved. However, he also stated that this was not a fundamental issue from an industry perspective. He suggested that less than 1 per cent of traded diamonds are linked to human rights violations. The representative also stated that further improvements were possible in relation to the Kimberley Process.[7]

---

5   Kimberley Process, *The Kimberley Process Certification Scheme* (Core Document, 2002) ss III, IV.
6   Interview with Global Witness Representative (telephone interview, 21 May 2007).
7   Interview with Rio Tinto Representative (telephone interview, 30 May 2007).

In order to improve the Kimberley Process, Global Witness and other organisations have urged that the process make explicit the internal controls that are required for different categories of countries. In particular, artisanal diamond producing and trading countries all need separate systems of controls, as do countries with cutting and polishing centres. The creation of the Artisanal Mining Working Group represents an important step in this direction.[8]

According to Global Witness, problems also exist at the point of retail in countries such as the United States. Global Witness argues that there is insufficient checking of shipments, because it is a low priority for business and government. In its interview response, Global Witness highlighted the issue of the ongoing sustainability of the Kimberley Process. With not as many diamonds fuelling conflict, the concern raised was that governments will give up on the Kimberley Process, thereby undermining its power as a preventative measure. As a result, there is the risk of renewed instability in artisanal diamond areas potentially leading to the outbreak of conflict.[9]

A further important mechanism of regulation is through the making and enforcement of contracts. Requirements regarding Kimberley Process obligations could be built into any potential contracts negotiated between government and business. For example, a government contract with a business enabling the mining of a concession could stipulate that the business guarantee that no revenue go to persons linked to human rights violations. Failure to ensure this stipulation could lead to the termination of the contract. Corporations entering into contracts with subsidiaries or other independent businesses could put guarantees of Kimberley Process compliance into their contracts. For example, a diamond retailer might put a guarantee that diamonds being purchased are conflict-free into their purchase contract from a diamond cutter/polisher. Failure to adhere to the

---

8   Global Witness, 'Kimberley Process Certification Scheme Questionnaire for the Review of the Scheme' (Review Submission, 5 April 2006). At its 6–9 November 2006 meeting, the Kimberley Process Plenary agreed to the creation of a Working Group on Artisanal-Alluvial Production. Kimberley Process, 'Kimberley Process Plenary: Final Communique' (Gaborone, Botswana, 6–9 November 2006).
9   Interview with Global Witness Representative (telephone interview, 21 May 2007).

obligation would constitute breach of the contract. It would be open to national governments to require that retailers or other businesses in the diamond supply chain include such provisions in their contracts.[10]

Regulatory Options Under National Legislation

Of importance and interest to any system of regulation is the ability of regulators to enforce their rules. The Kimberley Process core document is not silent simply on the issue of internal controls for governments, but also on the matter of how the import/export certification regime is to be enforced. National governments have an array of enforcement mechanisms available to them to ensure that national industry is compliant with the Kimberley Process. They are able to engage in negotiation and discussion with industry in order to promote compliance with the Kimberley Process. Their ability to engage in negotiation is backed up by the ability to impose sanctions for various types for non-compliance. Naturally, the first of such sanctions is the refusal to grant an export certificate under Kimberley Process procedures. Beyond this, national governments might impose financial penalties or even undertake criminal prosecutions for serious offences.

It would seem logical, in the implementation of the Kimberley Process on the national level, to provide for criminal sanctions for individuals who trade in diamonds in the absence of appropriate certification. This is the simplest method of providing a criminal sanction for violation of the Kimberley Process provisions, as all that must be proven is an absence of a valid Kimberley Process certificate. The sanction should have confiscation of the diamonds and some type of fine as a minimum penalty, with the option of terms of imprisonment where appropriate *mens rea* is present, including situations of repeat offenders. Where there is evidence that a person has been engaged in the conflict diamonds trade, as opposed to simply trading without a certificate, there are further options for national criminal prosecution. This might be considered a more serious offence, as it shows that the

---

10   Collins, H, 'Regulating Contract Law' in Christine Parker, et al. (eds), *Regulating Law* (Oxford University Press, 2004) 13–32.

attempt to export the diamonds was made in the knowledge that, at some point, the trade benefits militia groups involved in human rights abuses. Of arguably greater seriousness is the crime of contravening United Nations sanctions, where such sanctions have been imposed.[11]

One of the strengths of prosecutions for trafficking in conflict diamonds, or contravening United Nations sanctions, is the comparative simplicity of making out a case. Rather than requiring the judicial process to consider the intricacies of international crimes prosecutions, it is simpler for a court to consider evidence that an individual has been trading in conflict diamonds that are not correctly certified, and that the person possesses the requisite *mens rea*. Sanctions under national laws could also be applied to corporations themselves, an option not currently available under international criminal law.

Moving squarely into the province of serious international crimes is the prospect of prosecution domestically for conduct amounting to the war crime of pillage, as incorporated into domestic legislation by parties to the *Rome Statute of the International Criminal Court 1998*. Essentially, this crime involves the removal of resources and property without the consent of the legitimate government.[12]

A final option is prosecution for trading in conflict diamonds, by relying on indirect liability for the perpetration of war crimes and crimes against humanity such as murder and torture. Such an approach focuses on human rights violations committed by direct perpetrators, but attributes individual criminal responsibility to those whose role was financing the perpetrators through illegal diamond trading. Any country that is a party to the *Rome Statute* is obliged to legislate domestically for international crimes such as torture and

---

11   For example, see discussion of the Kouwenhoven case below, which involved a prosecution under Dutch laws that criminalised the infringement of United Nations sanctions.

12   *Rome Statute of the International Criminal Court*, opened for signature 17 July 1998, 2187 UNTS 90 (entered into force 1 July 2002) art 8(2)(b)(xvi) (prohibition of pillage in international conflicts), art 8(2)(e)(v) (for non-international conflicts). A further possibility is taking proceedings, based on state responsibility rather than individual criminal responsibility, before the International Court of Justice. Uganda was found to be in breach of its international obligation to prevent plunder by the International Court in the *Case Concerning Armed Activities on the Territory of the Congo (Democratic Republic of Congo v Uganda) (Judgement)* (1995) ICJ Rep 90 [213]–[214], [242]–[251]. The relevant obligation that had been breached was the rule concerning pillage in *Hague Convention (IV) Respecting the Laws and Customs of War on Land and Its Annex: Regulations Concerning the Laws and Customs of War on Land*, opened for signature 18 October 1907, (entered into force 26 January 1910) art 43.

murder. This possibility is the national analogue of an international prosecution before the court itself, and is discussed in the context of the Kouwenhoven case further below.

Beyond an in-principle consideration of domestic legislative approaches, the actual legislative framework from a number of diamond-producing countries in managing diamond mining and export, including provisions specifically directed against conflict diamonds, are discussed below. Most of the legislation considered relates to African diamond-producing nations. However, the legislative approach of a developed diamond-producing country, Australia, is also considered, by way of comparison. Finally, a domestic prosecution initiated by Dutch authorities in an analogous field to conflict diamonds, the Kouwenhoven case concerning so-called 'conflict timber', is discussed.

Angolan National Legislation

Angola has created a legal regime under its domestic legislation to manage diamond mining activities. The legislation employs two main approaches to the regulation of the mining industry, namely zoning and licensing.[13] Zoning involves the designation of certain areas as diamond production zones.[14] Public access to such areas is prohibited or restricted, and taking up residence in the area is regulated, as is the carrying on of certain economic activities there, and bringing particular goods into and out of the zone.[15] Criminal offences for contravention of these requirements are created.[16]

Under licensing laws, persons are authorised to carry out particular activities in relation to diamond mining activity, such as prospecting, exploration, extraction, trading, and, importantly, export

---

13   However, other regulatory approaches are also provided for. For example, an income tax regime for diamond mining is established under the *Regulamento do Regime Fiscal para a Indústria Mineira 1996* [Regulation of the Fiscal Regime for the Mining Industry] (Angola) Decree-Law No 4-B/96, 31 May 1996.

14   *Da Lei dos Diamantes 1994* [The Diamond Law] (Angola) Law No 1/92, 7 October 1994 arts 14–23; *Lei sobre o Regime Especial das Zonas de Reserva Diamantífera 1994* [Law on the Special Regime of Zones of Diamondiferous Deposits] (Angola) Law No 17/94, 7 October 1994.

15   *Da Lei dos Diamantes 1994* [The Diamond Law] (Angola) Law No 1/92, 7 October 1994 arts 14–23; *Lei sobre o Regime Especial das Zonas de Reserva Diamantífera 1994* [Law on the Special Regime of Zones of Diamondiferous Deposits] (Angola) Law No 17/94, 7 October 1994.

16   *Da Lei dos Diamantes 1994* [The Diamond Law] (Angola) Law No 1/92, 7 October 1994 arts 28–31.

of diamonds.[17] The government has the ability to inspect and audit all mining activities[18] and suspend or revoke mining licences if required.[19] Criminal offences exist for persons carrying out such activities in the absence of a licence.[20] Artisanal mining is specifically regulated through the use of particular zoning and licensing requirements.[21]

Angolan legislation provides, in particular, that diamond exporters must have an export licence, and sets out relevant customs procedures for exporters.[22] Classification and valuation of diamonds prior to export is regulated by the national diamond mining agency, Emprese Nacional de Diamontes.[23] Unauthorised importation of diamonds into the country is criminalised, as is unauthorised trading in diamonds.[24]

Congolese National Legislation

The DRC employs the tools of zoning and licensing to regulate its mining industry under its national legislative regime.[25] By the use of a zoning mechanism, the government may prohibit mining activity in particular areas.[26] When properly licensed, a natural person, public entity or corporation, whether foreign or Congolese, may carry out diamond mining on a large-scale basis.[27] Such activities include prospecting, exploration, extraction, trading and export of

---

17   *Lei das Actividades Geológicas e Mineiras 1992* [Law on Geological and Mining Activities] (Angola) Law No 1/92, 17 January 1992 arts 6–7. Articles 10–13 allow for the granting of exploration rights, a process of determining whether traces of a material are commercially exploitable. The legislation also provides for the licensing of extraction activities.

18   Ibid art 24.

19   Ibid arts 17, 22.

20   *Da Lei dos Diamantes 1994* [The Diamond Law] (Angola) Law No 1/92, 7 October 1994 ch 6, especially art 24.

21   Ibid arts 31–36 creates specific licensing and zone regulations for artisanal miners.

22   *Regime Aduaneiro Aplicável ao Sector Mineiro 1996* [Customs Regime Applicable to the Mining Sector] (Angola) Decree-Law No 12-B/96, 31 May 1996 arts 9, 10.

23   *Da Lei dos Diamantes 1994* [The Diamond Law] (Angola) Law No 1/92, 7 October 1994 art 10.

24   Ibid arts 36, 38; *Regime Aduaneiro Aplicável ao Sector Mineiro 1996* [Customs Regime Applicable to the Mining Sector] (Angola) Decree-Law No 12-B/96, 31 May 1996 arts 9, 10.

25   *Loi Portant Code Minier 2002* [Law Relating to the Mining Code] (The Democratic Republic of Congo) Law No 007/2002, 11 July 2002.

26   Ibid arts 6, 8.

27   Ibid art 5. Under this article, the applicant must specify the type of mineral that will be mined (e.g. diamonds).

diamonds.[28] Mining licence applications can be refused or revoked by the government.[29] Fines apply to persons undertaking unauthorised mining activities, with minerals being subject to confiscation.[30] Also criminalised are theft, possession of stolen minerals, unauthorised trading in minerals, and unauthorised transportation of minerals.[31] Fraudulent exports are also subject to criminal sanctions.[32]

A separate regime applies to artisanal miners. Such activities may be restricted to particular artisanal mining zones and may only be carried out by Congolese natural persons who are licensed for artisanal diamond mining.[33] Only authorised artisanal diamond traders may purchase diamonds from miners and sell them on the domestic market.[34] Only diamond trading houses, and their agents, may export artisanally mined diamonds.[35]

Côte d'Ivoire National Legislation

The Côte d'Ivoire legislative framework provides for protected zones where mining is not permitted, as well as zones where entry is not permitted without authorisation and zones for artisanal mining.[36] The legislation sets up licensing systems for mineral prospecting,

---

28 Ibid arts 18–21 (prospecting), chapter 1 (exploration), arts 63–66, 69, 73, 80 (exploitation, which includes extraction and sale of extracted minerals), art 85 (separate authorisation is required to export 'untreated ores', although it is unclear whether this applies to rough diamonds. Article 266 states that a licence holder can export and sell its production to international markets.). Articles 23 and 25 require that, apart from prospecting, foreign corporations must act through a mining agent in the county. The government requires 5 per cent of the shares of the registered capital of mining corporations who have an exploitation licence. Chapter III of the Mining Code provides that an exploitation licence may apply to artificial mineral deposits known as 'tailings'.
29 Ibid arts 27, 73, 80. Government officials are not permitted to be artisanal or large-scale miners or traders.
30 Ibid art 299.
31 Ibid arts 300–302.
32 Ibid art 234, which refers to sanctions located in customs legislation.
33 Ibid arts 5, 26–27, 109–118. A further category called 'small-scale mining' also exists under Chapter 5 of the Mining Code, which is larger than artisanal mining, but smaller than large-scale mining.
34 Ibid arts 27, 118–119.
35 Ibid arts 120–123. They are required to purchase all artisanally mined substances presented for sale, regardless of quantity or quality, under Article 123. Partnership Africa Canada, an NGO involved in the Kimberley Process, reported that there was evidence of bribes being taken by diamond valuers in the period since 2005, which was the year that the employment of an independent diamond valuer was discontinued. Partnership Africa Canada, 'Diamonds and Human Security: Annual Review 2008' (Annual Report, October 2008) 4–5.
36 *Code Minier 1995* [Mining Code] (Côte d'Ivoire) Law No 95-553, 17 July 1995 title VI (large-scale mining), art 42 (artisanal mining).

exploration and exploitation, as well as artisanal mining.[37] Furthermore, the legislation establishes criminal offences to enforce the licensing system, which are punishable by fines, imprisonment or both. Such offences include exploitation, possession, trade and transportation of minerals without authorisation, as well as fraud in relation to these activities.[38]

Sierra Leone National Legislation

The Sierra Leone legislative framework also provides for zoning in relation to artisanal mining activities.[39] Artisanal miners, traders and exporters must also be licensed, and artisanal diamonds are only allowed to be traded between such persons, unless exported under a valid export licence.[40] Large-scale mining activities, including prospecting, exploration, exploitation and export are also subject to a licensing regime.[41] Fines and terms of imprisonment may be imposed if a criminal offence is successfully prosecuted. Offences include mining, possession, sale or export of minerals without authorisation.[42]

Australian National Legislation

The Kimberley requirements for the export of rough diamonds were implemented under Australian domestic law under Regulation 9AA of the Customs (Prohibited Exports) Regulations 1958. Under this regulation, the responsible minister may, on application, grant permission for the exportation of rough diamonds to a country by issuing a certificate. The regulation prohibits the export of rough diamonds unless the exporter holds a Kimberley Process certificate, the original is produced to customs at or before the time of exportation and the rough diamonds are exported in a tamper-resistant container.[43]

---

37   Ibid titles II, III (large-scale mining), arts 43–45 (artisanal mining).
38   Ibid arts 101–108.
39   *Mines and Minerals Act 1994* (Sierra, Leone) s 20.
40   Ibid ss 76, 79–81, 118.
41   Ibid ss 1, 49, 54–55, 61, 67, 118.
42   Ibid ss 117–118. The legislation also provides for a number of revenue-raising mechanisms for the government (Part XIII).
43   Written Response from Australian Government to Author's Interview Questions (14 September 2007).

Australia has a two-tier system of managing export certification. Occasional exporters are required to apply to the relevant federal department in writing for each shipment and provide documentary evidence of the origin of rough diamonds they wish to export. Applicants are required to complete a criminal history check through the Australian Federal Police. Packages of rough diamonds to be exported must be declared to customs on departure and the certificate presented.[44]

Businesses wishing to regularly export shipments of rough diamonds may apply for a frequent exporter licence. Successful applicants are pre-issued with sequentially numbered stocks of partially completed Kimberley Process certificates. Using database applications supplied by the relevant federal department, the companies are able to complete the details of the goods for shipment and the importing business and then finalise the validation of the Kimberley Process certificate. Details of each Kimberley Process certificate are transmitted to the relevant federal department and incorporated into the Export Authority's database records. Under this decentralised system, frequent exporters work closely with Australian Kimberley Process authorities to manage certification procedures. Customs verifies the shipment satisfies the requirements of regulations at export.[45]

Under import regulation, Regulation 4MA of the Customs (Prohibited Imports) Regulations 1956, the import of rough diamonds is prohibited unless the diamonds are accompanied by a Kimberley Process certificate, they are imported in a tamper-resistant container and the diamonds are imported from a country that is a participant in the Kimberley Process.[46]

The *Customs Act 1901 (C'th)* provides customs with the appropriate powers to enforce compliance with the regulations and take appropriate action. Shipments not meeting the requirements of import or export controls may be detained or seized by customs. Customs undertakes checks to verify compliance in an environment that is largely self-regulated. Customs intervenes in transactions proportionate to the perceived level of risk.[47]

---

44   Ibid.
45   Ibid.
46   Ibid.
47   Ibid.

The intervention of customs is generally aimed at encouraging compliance and is appropriate to the assessed level of risk. Customs compliance programs focus on assisting clients who are willing and capable of complying with the legislation but there is scope to impose sanctions on entities where appropriate.[48] Cooperation with industry, facilitated by the relatively concentrated nature of diamond mining industry in Australia has been one of the major strengths of Australia's implementation of the Kimberley Process.[49]

For exports of diamonds from Australia, the relevant federal department may decline to grant a Kimberley Process certificate where the circumstances warrant such action. Import or export of prohibited diamonds is an offence under the *Customs Act 1901*. Regulation 9AA of the Export Regulations makes it an offence to export rough diamonds from Australia without a Kimberley Process certificate. Regulation 4MA of the Customs Import Regulations makes it an offence to import rough diamonds into Australia without a Kimberley Process certificate.

Regulation 4N of the Import Regulations and Regulation 8A of the Charter of the United Nations (Sanctions – Côte d'Ivoire) make it an offence to import rough diamonds into Australia from Côte d'Ivoire (whether or not the rough diamonds originated in Côte d'Ivoire) or of Côte d'Ivoire origin from a third country.

In all cases, offences only apply where rough diamonds are exported from or imported into Australia. There is no specific offence under Australian law of trading in rough diamonds where the trade does not involve Australian territory. The penalty on conviction is not greater than three times the value of the goods or AUD$110,000 whichever is greater.

According to the Global Witness interview comments, Australia has not had a review visit and there has been no independent verification of its Kimberley Process controls. However, Australia has submitted annual reports every year, and appears to comply with minimum requirements. Global Witness noted that there is generally not much of an issue with conflict diamonds in the large-scale industrial mining sector that is present in Australia. However, the respondent stated

---

48   Ibid.
49   Ibid.

that Australia appeared to have hands-off regulation much like the US system, and there was a question as to what checks are going on and whether there is sufficient government oversight.[50]

Dutch National Legislation and the Kouwenhoven Case

As discussed above, the utilisation of national war crimes legislation represents a possible avenue for national governments to bring to account those engaged in the conflict diamonds trade. The Netherlands utilised its national war crimes legislation to initiate a prosecution about the related issue of so-called 'conflict timber'. Although not a conflict diamonds prosecution as such, the war crimes legislation was used to prosecute timber trader and Dutch national, Gus Kouwenhoven. Reminiscent of the conflict diamonds problem, Kouwenhoven allegedly provided financial assistance through his logging activities to human rights violators.[51] Kouwenhoven was charged with war crimes for his role in the conflict in Liberia, as well as breaching United Nations sanctions.

The indictment alleged that in at least four locations, Kouwenhoven committed, directly or indirectly, the war crimes of killing, inhuman treatment, looting, rape, severe bodily harm, and offences against dead, sick or wounded persons.[52] Machine guns and rocket-propelled grenades were used in an attack that made no distinction between

---

50   Interview with Global Witness Representative (telephone interview, 21 May 2007).

51   Global Witness, *Natural Resources and Conflict under the Legal Spotlight, War Crimes Trial of Gus Kouwenhoven to Commence in The Hague* (21 April 2006). Available at: www.globalwitness.org/library/natural-resources-and-conflict-under-legal-spotlight-war-crimes-trial-gus-kouwenhoven. An interesting discussion of the illegal timber trade and its parallels with the conflict diamonds trade and the Kimberley Process can be found in Salo, Rudy S, 'When the Logs Roll Over: The Need for an International Convention Criminalizing Involvement in the Global Illegal Timber Trade' (2003) 16 *Georgetown International Environmental Law Review* 127, 127–146.

52   *Guus Kouwenhoven, Rechtbank 's-Gravenhage* [District Court of The Hague], Case No AY5160, 7 June 2006, 3. The international criminal law provisions allegedly contravened were the international customary law prohibition on indiscriminate attacks, as well as contraventions of *Geneva Convention Relative to the Treatment of Prisoners of War (Geneva Convention III)*, opened for signature 12 August 1949, 75 UNTS 135 (entered into force 21 October 1950) art 130 and common article 3 of the Geneva Conventions. The latter two provisions prohibit torture, inhuman treatment, rape, looting and acts of violence against inactive soldiers and civilians. The modes of individual criminal responsibility with which Kouwenhoven was charged were direct commission, aiding and abetting and superior responsibility.

active combatants and civilians.[53] In addition, a house was set on fire, resulting in the deaths of civilians and inactive soldiers.[54] Such persons were also killed in buildings through the use of grenades.[55] Civilians and inactive soldiers had their hands amputated, and babies were injured.[56] Women and children were raped, and the possessions of civilians and inactive soldiers were plundered.[57]

Kouwenhoven's alleged role in these crimes was through selling or supplying weapons, vehicles and equipment, such as machine guns, rocket-propelled grenades, helicopters and trucks to Charles Taylor and his armed forces, and placing staff members with timber companies under threat of dismissal from employment at the disposal of the armed conflict. He was also charged with supplying or giving money, cigarettes, and marijuana to members of the armed forces of Charles Taylor, the Liberian Government, and timber company employees assisting the conflict. He allegedly gave instructions regarding the use of weapons, including heavy weapons and battle methods to the armed forces of Charles Taylor, the government of Liberia and staff members assisting the conflict, telling them that all should be killed without distinction and that looting was accepted.

In two further counts, Kouwenhoven was charged with contravening United Nations sanctions on Liberia, criminalised under domestic Dutch legislation. In particular, Kouwenhoven was charged with supplying Taylor's forces with machine guns, grenade launchers, and mortars.[58]

The decision of the District Court of The Hague in the Kouwenhoven case was delivered on 7 June 2006.[59] The court found that these basic crimes were committed. However, concerning the role of the accused, the court did not find that Kouwenhoven was individually criminally responsible. In particular, the court found there was insufficient evidence to prove that the defendant had knowledge of the alleged criminal activities of his subordinates. In most instances the security

---

53   *Guus Kouwenhoven, Rechtbank 's-Gravenhage* [District Court of The Hague], Case No AY5160, 7 June 2006, 4.
54   Ibid.
55   Ibid.
56   Ibid.
57   Ibid.
58   Ibid.
59   Ibid.

employees were former fighters of Charles Taylor's armed forces, and so it could not be concluded that the security employees participated in these acts by order of or with the consent or knowledge of the defendant. The court found it proven that the defendant, together and in conjunction with another, supplied weapons to Charles Taylor, but found that this was not itself sufficient evidence that he directly participated in committing the offences charged under the three counts. The court noted that weapons could have been used for legal acts.

The court, however, found Kouwenhoven guilty of charges four and five, relating to the contravention of United Nations sanctions. Based on evidence adduced in court, the court found that there was a close financial relationship between the defendant and Taylor. There were also personal ties between Kouwenhoven's timber company staff and Liberian Government members. The court found it established that Kouwenhoven played an important role in supplying weapons to Taylor and Liberia. In reaching this conclusion, the court relied in particular on the fact that Kouwenhoven was the owner of a ship called the *Antarctic Mariner*, which had been used to import weapons for the benefit of Taylor and his regime. From this, the court concluded that in 2000, Kouwenhoven was involved in the contravention of UN sanctions. The court rejected a defence to weapons importation on the grounds that the weapons were needed as legitimate self-defence by Liberia. Instead, the court found that the United Nations Security Council had already pronounced that Liberia was acting illegally through supporting the Revolutionary United Front rebels based in Sierra Leone.[60]

The conviction of Kouwenhoven at first instance was overturned on appeal on 10 March 2008.[61] According to the Court of Appeal, the written testimony of a number of witnesses was not considered sufficiently reliable to ground a finding connecting Kouwenhoven with the transport of weapons.

---

60   Ibid.
61   British Broadcasting Corporation, 'Profile: Guus van Kouwenhoven' (2008). Available at: news.bbc.co.uk/2/hi/africa/5055442.stm.

Although there was no conviction recorded against Kouwenhoven as a result of insufficient evidence, the case is an important precedent for war crimes prosecutions where an individual has allegedly assisted the crime financially rather than being a direct perpetrator. The court was clearly quite prepared to consider the allegations that Kouwenhoven contributed to the commission of the crimes through such indirect means as providing financial assistance, weapons, training, and enforced enlistment of employees to fight in the conflict. If these modes of involvement are recognised as being sufficient in principle to bring about a war crimes conviction, then it would appear that trading in conflict diamonds, to the benefit of human rights violators, and with the appropriate *mens rea,* may also be sufficient to lead to a guilty verdict in a domestic war crimes case.

A further development occurred in 2010, when the Dutch Supreme Court overturned the acquittal by the Appeal Court. The trial re-opened in December 2010, and Kouwenhoven still faces charges of war crimes and illegal arms trading. In November 2014, the court in Den Bosch was to hear arguments for dismissal as there were no witnesses available to testify. The Prosecutor Cara Pronk-Jordan wishes to have the earlier anonymous interviews used as evidence.[62]

Concluding Remarks

The chapter discussed the implementation of the Kimberley Process at the national level, including regulatory options under national legislation in Angola, the DRC, Côte d'Ivoire, Sierra Leone, Australia, and the Netherlands. A review of legislation and regulatory action at the national level is necessary in terms of responding to the first research question: to what extent has the conflict diamonds governance system achieved its objectives? Legislation in African diamond-producing states focuses on zoning and licensing systems. Zoning is a method of distinguishing kimberlite, industrial areas of mining from artisanal-alluvial mining areas, and is used to designate areas for diamond mining exploration. Licensing denotes systems of authorising persons to carry out particular diamond mining–related

---

62   Global Witness, *Global Witness Welcomes Dutch Court's Decision to Hear New Prosecution witnesses in Kouwenhoven Case.* Available at: www.globalwitness.org/en/archive/global-witness-welcomes-dutch-courts-decision-hear-new-prosecution-witnesses-kouwenhoven/.

activities, such as prospecting, exploration, kimberlite mining or artisanal mining. Both zoning and licensing are potentially useful regulatory techniques, with the caveat that implementation of the law is carried out within the bounds of international human rights law. Zoning areas for artisanal mining potentially allows for these miners to carry on a livelihood, while protecting the interests of large-scale industry that are focused more on kimberlite deposits. However, enforcement of zoning, namely moving artisanal miners from kimberlite areas, or unauthorised artisanal miners, has become a justification for government security forces to carry out serious human rights abuses such as rape and murder in countries such as the DRC, Zimbabwe, and Angola. Similarly, the use of licensing, particularly for artisanal miners, as well as diamond traders and exporters, if effectively monitored and enforced, would represent an effective internal control set to ensure that all diamonds being mined and exported are legitimate rather than funding rebel militias or the commission of international crime. The challenge with licensing thus relates to both excessive and inadequate enforcement. Angola and Zimbabwe have seen the excessive use of force, resulting in murder, beatings and rape, in the name of enforcement of rights to mine in artisanal mining areas. At the other end of the spectrum are artisanal areas where there are inadequate bureaucratic resources to manage licensing systems correctly, or in which corruption means that the licensing system will not work properly.

The Australian example shows that Kimberley Process certification requirements, for both import and export, have been implemented into domestic legislation. The challenge of applying the Australian legislative provisions to those of the African producer states lies in the fact that diamond mining in these African countries is largely artisanal, whereas diamond mining in Australia is solely of industrial, kimberlite deposits. A further difference is that Australian diamonds are predominantly industrial grade, and are used as drill bits, rather than jewellery pieces. The Australian approach could therefore be used in relation to large-scale kimberlite mining, for example, in Angola, which has both kimberlite and artisanal sites, but is of little value to the informal artisanal sector. For example, Australia gives blanket approval and Kimberley Process certification in advance to the Rio Tinto operation, although it provides for inspections of the Rio Tinto operation to ensure that domestic regulatory requirements are being adhered to. It is possible to argue that this approach is appropriate to a large-scale operation in a country not experiencing civil war, and where there are no known international crimes being committed in connection with the mine. Australia takes a different approach for small packets of diamonds that are being exported: the Kimberley Process certificate would only be available by application to the department. Naturally, rough diamond imports require Kimberley Process certificates before importation is accepted.

The Australian experience, on reflection, may not be a good model even in relation to kimberlite diamond mining in African producer countries, because countries such as Angola are only just emerging from situations of civil war, and so a more careful approach to issuing Kimberley Process certificates is warranted. This is particularly the case in Angola, where military and police forces have been implicated in gross human rights abuses since the end of the civil war period. As such, allowing corporations to self-administer KP certificates, even in the presence of regular inspections, would not be a sensible way forward. The Australian experience does not throw light onto the administration of KP internal controls in the artisanal industry. As previously discussed, a system of licensing in principle would be effective, but would require efficient implementation by a bureaucracy that is free from corruption.

The chapter has also considered the landmark Kouwenhoven case, prosecuted in the Netherlands court system. Although it was not a conflict diamonds case, it was a case concerning the interaction between resources, in this case the timber industry, and international human rights crimes committed during the course of the Liberian civil war. An interesting side issue is that this case concerns the relationship of Kouwenhoven to the Liberian regime of Charles Taylor, who was a key player in conflict diamonds trafficking. However, the central value of the case lies in the fact that it largely parallels the type of prosecution that could be initiated in relation to conflict diamonds trading. Instead of trading in diamonds, thereby funding the purchase of weapons and the commission of human rights abuses, Kouwenhoven was accused of misusing his management of the timber industry to assist Charles Taylor's army to carry out gross human rights violations. His timber corporation was allegedly used as a means of purchasing armaments that were later deployed by Taylor's forces. Kouwenhoven allegedly directed his timber corporation employees to join Taylor's forces, which then carried out human rights violations, including the murder of civilians. The case is yet to be finally resolved by the Dutch court system as the acquittal was overturned in 2010, and the case reopened to hear new prosecution evidence from December of that year.

What is perhaps most interesting about the Kouwenhoven case is that there was no debate, as a matter of law, as to Kouwenhoven's criminal liability, with the multiple appeals turning on whether the evidence was credible and able to sustain the conviction. This means that, regardless of the final outcome of the case, the Dutch court system has accepted that, should it be proved that a person, for example, forced their employees to join an armed force that went on to commit international crimes, that person could be found to be legally responsible for the crimes committed. In this way, the case shows that business leaders can be convicted for their role in the commission of international crimes. It is furthermore possible to imagine that some of the ways that Kouwenhoven was involved in the commission of crimes, through his connection with a timber corporation, might be applied to a different business leader who is involved in the traffic of rough diamonds rather than timber.

5

Growing Teeth: The UN Security Council and International Tribunals

What had meaning in this conflict were diamonds. Between 1998 and 2000, diamonds mined by forced labour were first taken to the headquarters in Buedu and from there to the accused [Charles Taylor] in Liberia. In return ... arms were then distributed to the AFRC/RUF forces ...

Prosecutor's opening statement, Taylor case[1]

Chapter Overview

This chapter looks beyond the Kimberley Process (KP) to the other international organisations that contribute to the conflict diamonds governance system. After considering briefly the role of the United Nations General Assembly (UNGA), I consider the role of the United Nations Security Council (UNSC) in creating the KP and enhancing it through its monitoring and enforcement activities. Finally, the chapter considers the jurisprudence emerging from the Special Court for Sierra Leone and the International Criminal Court (ICC) about conflict diamonds in the context of the commission of international human rights crimes. The jurisprudence uses conflict diamonds in three major ways: as the context in which international crimes are committed, as

---

1   Transcript of Proceedings, *The Prosecutor v Charles Taylor (Trial)*, (Special Court for Sierra Leone, Trial Chamber II, Case No SCSL-03-01-PT, 4 June 2007) 59.

a direct aspect of an international crime, and as indirectly imposing liability for the international crimes committed by others. The chapter's concluding remarks highlight the important relationship between international human rights crimes, the KP, and the concept of conflict diamonds.

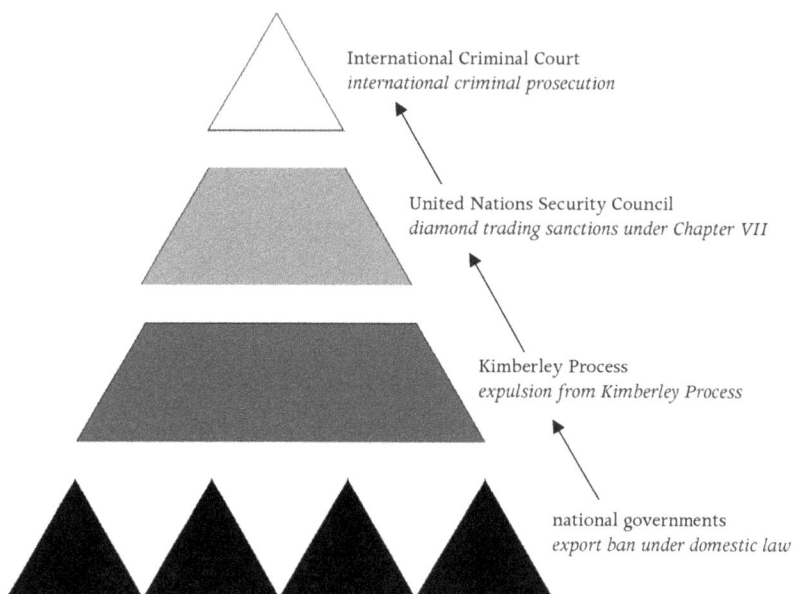

Figure 5.1: Conflict Diamonds Governance as a Networked Pyramid of Sanctions
Source: Author's research.

United Nations General Assembly

The UNGA has played a significant role in the establishment of the Kimberley Process and in providing it with ongoing political support. The Kimberley Process was initially called for under United Nations General Assembly Resolution 55/56. The legitimacy and authority of the Kimberley Process has subsequently been reinforced on an annual basis by UN General Assembly and Security Council resolutions. It has been noted, however, that, unlike a formal mandate under the coercive powers of the Security Council, General Assembly resolutions are not legally binding. The three-year review of the Kimberley Process Certification Scheme (KP) called for the UN Secretariat to be regularly invited to KP Plenary meetings and kept informed of KP decisions. It is interesting to note that although the KP exists as a non-legally binding, informal institution, it is largely a product of the UN system, which continues to monitor its activities, provide political support and, in the case of the UNSC, intervene in particular enforcement scenarios.[2]

United Nations Security Council

Prior to the establishment of the Kimberley Process, the Security Council played a key role in bringing to international attention the problem of conflict diamonds. In 1998, it adopted resolutions 1173 and

2   SC Res 827, UN SCOR, 3217th mtg, UN Doc S/RES/827 (25 May 1993) ('Statute of the International Criminal Tribunal for the Former Yugoslavia'); The Role of Diamonds in Fuelling Conflict: Breaking the Link Between the Illicit Transaction of Rough Diamonds and Armed Conflict as a Contribution to Prevention and Settlement of Conflicts GA Res 55/56, UN GAOR, 55th sess, 79th plen mtg, A/RES/55/56 (1 December 2001); The Role of Diamonds in Fuelling Conflict: Breaking the Link Between the Illicit Transaction of Rough Diamonds and Armed Conflict as a Contribution to Prevention and Settlement of Conflicts, GA Res 56/263, UN GAOR, 56th sess, 96th plen mtg, UN Doc A/Res/56/263 (13 March 2002); The Role of Diamonds in Fuelling Conflict: Breaking the Link Between the Illicit Transaction of Rough Diamonds and Armed Conflict as a Contribution to Prevention and Settlement of Conflicts GA Res 57/302, UN GAOR, 57th sess, 83rd plen mtg, UN Doc A/RES/57/302 (15 April 2003); The Role of Diamonds in Fuelling Conflict: Breaking the Link Between the Illicit Transaction of Rough Diamonds and Armed Conflict as a Contribution to Prevention and Settlement of Conflicts GA Res 58/290, UN GAOR, 58th sess, 85th plen mtg, UN Doc A/RES/58/290 (14 April 2004); The Role of Diamonds in Fuelling Conflict: Breaking the Link Between the Illicit Transaction of Rough Diamonds and Armed Conflict as a Contribution to Prevention and Settlement of Conflicts GA Res 60/182, UN GAOR, 60th sess, 67th plen mtg, UN Doc A/RES/60/182 (20 December 2005); The Role of Diamonds in Fuelling Conflict: Breaking the Link Between the Illicit Transaction of Rough Diamonds and Armed Conflict as a Contribution to Prevention and Settlement of Conflicts GA Res 61/28, UN GAOR, 61st sess, 64th plen mtg, UN Doc A/RES/61/28 (4 December 2006); Report of the Kimberley Process Certification Scheme to the General Assembly pursuant to resolution 61/28, by Fernando M. Valenzuela on behalf of Kimberley Process Chair, 62nd sess, Agenda Item 13, UN Doc A/62/543 (13 November 2007); The Role of Diamonds in Fuelling Conflict: Breaking the Link Between the Illicit Transaction of Rough Diamonds and Armed Conflict as a Contribution to Prevention and Settlement of Conflicts, 62nd sess, Agenda Item 13, UN Doc A/62/L.16 (21 November 2007); The Role of Diamonds in Fuelling Conflict: Breaking the Link Between the Illicit Transaction of Rough Diamonds and Armed Conflict as a Contribution to Prevention and Settlement of Conflicts, 63rd sess, Agenda Item 11, UN Doc A/63/L.52 (5 December 2008); The Role of Diamonds in Fuelling Conflict: Statement to General Assembly, Kimberley Process Chair Nirupam Sen, 63rd sess, 67th plen mtg, Agenda Item 11, UN Doc A/63/PV.67 (11 December 2008). See also, Written Response from Australian Government to Author's Interview Questions (14 September 2007).

1176, prohibiting the direct or indirect export of unofficial Angolan diamonds.[3] In 1999, the UNSC passed a resolution prohibiting trading in diamonds originating from conflict-ridden Sierra Leone.[4]

In regulatory terms, the resolutions are important examples of the standard-setting and enforcement roles of the Security Council. The legally binding nature of these resolutions means that the rulings must be complied with by all nations. The imposition of diamond trading bans represents an important mechanism for breaking the link between the illegal diamond trade and the commission of serious human rights violations.

The Security Council has engaged with the issue of conflict diamonds over a number of years, and has intervened decisively with resolutions concerning Angola, Sierra Leone, Liberia, the Democratic Republic of Congo (DRC), and Côte d'Ivoire. At these times, the Security Council has mandated trading bans with these countries concerning the rough diamond trade.[5]

The main mechanism that the Security Council has made use of in terms of monitoring has been its panel of experts. The panel of experts mechanism is important, as it provides vital on-the-ground fact-finding that can be used to assess the requirements for further international action, including the relative success of already existing mechanisms.

---

3   SC Res 1173, UN SCOR, 3891st mtg, S/RES/1173 (12 June 1998) (Angola); SC Res 1176, UN SCOR, 3894th mtg, UN Doc S/RES/1176 (24 June 1998). See also Interview with Global Witness Representative (telephone interview, 21 May 2007).
4   SC Res 1306, UN SCOR, 4168th mtg, S/RES/1306 (5 July 2000) (Sierra Leone).
5   SC Res 864, UN SCOR, 3277th mtg, S/RES/864 (5 June 1998) (Angola); SC Res 1127, 3814th mtg, S/RES/1127 (28 August 1997) (Angola); SC Res 1171, UN SCOR, 3889th mtg, UN Doc S/RES/1171 (5 June 1998) (Sierra Leone); SC Res 1173, UN SCOR, 3891st mtg, S/RES/1173 (12 June 1998) (Angola); SC Res 1295, UN SCOR, 4129th mtg, UN Doc S/RES/1295 (18 April 2000) (Angola); SC Res 1306, UN SCOR, 4168th mtg, S/RES/1306 (5 July 2000) (Sierra Leone); SC Res 1343, UN SCOR, 4287th mtg, S/RES/1343 (7 March, 2001) (Liberia); SC Res 1385, UN SCOR, 4442nd mtg, UN Doc S/RES/1385 (19 December 2001) (Sierra Leone); SC Res 1408, UN SCOR, 4256th mtg, UN Doc S/RES/1408 (6 May 2002) (Liberia); SC Res 1457, UN SCOR, 4691st mtg, S/RES/1457 (24 January 2003) (Democratic Republic of Congo); SC Res 1459, UN SCOR, 6494th mtg, UN Doc S/RES/1459 (28 January 2003) (Kimberley Process Certification Scheme); SC Res 1521, UN SCOR, 4890th mtg, UN Doc S/RES/1521 (22 December 2003) (Liberia); SC Res 1579, UN SCOR, 5105th mtg, UN Doc S/RES/1579 (21 December 2004) (Liberia); SC Res 1643, UN SCOR, 5327th mtg, S/RES/1643 (15 December 2005) (Côte d'Ivoire).

In reports as early as 1997, such expert panels have assessed the impact of the diamond trade in conflict zones, as well as the efficacy of diamond trade prohibitions and other economic sanctions.[6]

Since the establishment of the Kimberley Process, the Security Council has operated more closely with this new international agency. A particularly noteworthy example relates to the situation of Côte d'Ivoire. In 2006 it came to the attention of the Kimberley Process that conflict diamonds were propping up the rebel-held north of Côte d'Ivoire. Following an inspection by a KP monitoring team, the KP took action to expel Côte d'Ivoire from the KP. Shortly after, the matter was considered by the UNSC, which imposed a ban on trading in diamonds with that country.[7] The report on developments in Côte d'Ivoire referred to under Resolution 1572 was a joint mission undertaken in April 2006.[8]

There was also close cooperation with the UN Group of Experts appointed to report on developments in Liberia under UNSC Resolution 1521. By 2007, Liberia had notably made a transition out of its situation of conflict, had held internationally recognised elections, and elected a new president, Joan Sirleaf. In March of that year, a Kimberley Process review mission led by the European Community concluded that Liberia had designed effective controls in line with KP requirements. Although there were reservations from the NGO participants that Liberia would benefit from more time to ensure its

---

6   Progress Report of the Secretary-General on the United Nations Observer Mission in Angola (MONUA), S/1997/640 (13 August 1997); Report of the Panel of Experts on Violations of Security Council Sanctions Against UNITA, S/2000/203 (10 March 2000); Report of the Panel of Experts Appointed Pursuant to Security Council Resolution 1306 (2000), paragraph 19, in Relation to Sierra Leone, S/2000/1195 (20 December 2000); Report of the Panel of Experts on the Illegal Exploitation of Natural Resources and Other Forms of Wealth of the Democratic Republic of Congo, S/2001/357 (12 April 2001); Supplementary Report of the Monitoring Mechanism on Sanctions Against UNITA, 2001/966 (12 October 2001); Final Report Panel of Experts on the Illegal Exploitation of Natural Resources and Other Forms of Wealth of the Democratic Republic of the Congo, S/2002/1146 (16 October 2002); Report of the Group of Experts Submitted Pursuant to Paragraph 7 of Security Council Resolution 1584 (2005) Concerning Côte d'Ivoire, S/2005/699 (7 November 2005); Update Report of the Group of Experts Submitted Pursuant to Paragraph 2 of Security Council Resolution 1632 (2005) Concerning Côte d'Ivoire, S/2006/204 (31 March 2006); Interim Report of the Group of Experts on the Democratic Republic of the Congo, Pursuant to Security Council Resolution 1698 (2006), S/2007/40 (31 January 2007).
7   See also Interview with Global Witness Representative (telephone interview, 21 May 2007), with the representative commenting favourably on this coordination, although she noted that the action of the Security Council should have occurred more rapidly.
8   Written Response from Australian Government to Author's Interview Questions (14 September 2007).

resilience against the return of conflict diamonds, the KP presented its findings to the UNSC. The UNSC decided to lift its diamond embargo on Liberia on 27 April 2007 and Liberia was admitted to the KP on 4 May 2007.[9]

One of the important strengths of the Security Council resolutions is the legitimating role it plays at the apex of the United Nations security system. For example, the Security Council resolutions have enhanced the support base for the Kimberley Process, with significant momentum being gained for US national implementation legislation as a result of Security Council backing. The resolutions are, to an extent, self-contained mechanisms, as national governments are legally obliged to take action to implement them, even in the absence of national implementing legislation or involvement in a scheme such as the Kimberley Process.

Security Council expert panel reports have themselves acted as a type of enforcement mechanism, in the sense of naming and shaming particular individuals and corporations who have allegedly been involved in the conflict diamonds trade. Following its 2002 report on the DRC, the expert panel engaged in a negotiated process by which corporate and individual names could be removed from the black list following evidence of significant compliance with United Nations resolutions. The expert panel also reported a number of corporations to the Organisation for Economic Co-operation and Development (OECD) under the OECD multinational code of conduct for further action. Under this code, multinationals are required to account for their conduct before a committee. The OECD process, however, has been criticised as lacking in significant punitive powers in the event that breaches are found to have occurred. One approach that wasn't apparently considered, but was open to the Security Council, was the referral of individuals, including corporate directors, for prosecution by the ICC. This contingency is discussed in more depth below.

The consciousness-raising and information-sharing role of the Security Council has continued in more recent times. On 25 June 2007 there was a thematic debate at the UNSC following a proposal by the Belgian Mission on the theme 'Natural Resources and Armed Conflict', involving amongst other themes the illicit trade in diamonds.

---

9   Ibid; Interview with Global Witness Representative (telephone interview, 21 May 2007).

The council adopted a presidential statement on natural resources and armed conflict that underlined the importance of taking this dimension into account, where appropriate in the mandates of UN and regional peacekeeping operations, within their capabilities.[10]

The Security Council stated that natural resources are a crucial factor in contributing to long-term economic growth and sustainable development, while noting that in armed conflict situations the exploitation of natural resources has played a role in the outbreak, escalation, or continuation of the conflict. The council emphasised the contribution of monitoring schemes such as the KP.[11]

The response to the crisis in Côte d'Ivoire also illustrates the high level of coordination between the Kimberley Process and the UNSC. The Working Group on Monitoring monitored the situation regarding illicit mining in Côte d'Ivoire throughout 2005, drawing on a variety of reliable sources, and cooperated closely with the UN Panel of Experts on Côte d'Ivoire, which was convened by the Security Council. The Kimberley Process Secretariat then made a formal submission to the UN panel in June 2005, while a special envoy to the chair visited Abidjan in April 2005 to clarify Côte d'Ivoire's status in the Kimberley Process.[12] A special working group to deal with conflict diamonds from Côte d'Ivoire was established by the Kimberley Process at the 15–17 November plenary meeting.[13] On the basis of a presentation by the working group of the evidence available regarding production in Côte d'Ivoire, the Moscow Plenary meeting of November 2005 adopted a comprehensive package of measures to tackle the outflow of conflict diamonds from Côte d'Ivoire. Implementation of these measures is proceeding under the responsibility of the chair assisted by the Working Group on Monitoring.[14] Despite the rapid response of the working group to the crisis, and its cooperation as reflected in United Nations Security Council Resolution 1643 (2005), the working group

10   Written Response from Australian Government to Author's Interview Questions (14 September 2007).
11   Ibid.
12   Kimberley Process Working Group on Monitoring, *Submission for the 2006 Review of the KPCS* (Kimberley Process Secretariat, February 2006) 19–20.
13   Update Report of the Group of Experts Submitted Pursuant to Paragraph 2 of Security Council Resolution 1632 (2005) Concerning Côte d'Ivoire, UN Doc S/2006/204 (31 March 2006) 10.
14   Kimberley Process Working Group on Monitoring, *Submission for the 2006 Review of the KPCS* (Kimberley Process Secretariat, February 2006) 19–20.

recognised that the Kimberley Process's action has not by itself put an end to the illicit trade in Côte d'Ivoire conflict diamonds, and that further action is required.[15]

International Criminal Tribunals

Holding individuals criminally liable for particular conduct not only involves serious sanction for that person, but also sends a strong message to the international community. The liability for international crimes of those involved in industry was recognised at the Nuremberg trials of Farben, Flick, and Krupp following World War Two.[16] Recently established international courts and tribunals now provide the legal mechanisms for prosecuting modern industrial crimes, with a number of current conflict diamonds cases setting important international precedents.

Internationally, the UNSC has acted as a mediator between regulatory bodies such as the Kimberley Process and international criminal processes. One option that it has exercised has been the creation of ad hoc international criminal tribunals, such as the Sierra Leone Special Court, for the prosecution of international crimes that involve a conflict diamonds dimension. The court, which is a hybrid institution jointly established by the United Nations and the Sierra Leone Government, was established to prosecute international crimes committed in the territory of Sierra Leone since 1996. Due to the conflict diamonds dimension of the Sierra Leone conflict, the special court is uniquely placed amongst the ad hoc international criminal tribunals to carry out international conflict diamonds prosecutions.[17]

---

15   Ibid 20.
16   *Trials of Nazi War Criminals before the Neurnberg Tribunals under Control Council Law No 10: Neurnberg October 1946–April 1949,* Volumes V, VI, VII, VIII, IX (United States Government Printing Office, 1952). Available at: www.mazal.org/NMT-HOME.htm. Nuremberg prosecutors give useful commentary on the trials in these works. Taylor, T, 'Final Report to the Secretary of the Army on the Nuernberg War Crimes Trials Under Control Council Law No. 10' (William S. Hein & Co: Buffalo, 1997) 184–202; Sprecher, D A, *Inside the Nuremberg Trial: A Prosecutor's Comprehensive Account,* Vol. 1 (University Press of America, 1999) 134–137.
17   Website of the Residual Special Court for Sierra Leone. Available at: www.rscsl.org.

Sankoh, leader of the Revolutionary United Front (RUF), was arrested in 2000, and was to have been put on trial as part of the first case before the special court. Although Sankoh died before the commencement of his trial in 2004, proceedings commenced against other high-level members of the RUF, namely Issa Hassan Sesay, Morris Kallon, and Augustine Gbao.[18] The Trial Chamber delivered its judgement, convicting all three, in this combined case on 2 March 2009, and the Appeals Chamber confirmed the convictions on 26 October 2009.

In June 2003, the Sierra Leone Special Court unsealed an indictment against Liberian President Charles Taylor in relation to his alleged involvement in violations of international criminal law during the Sierra Leone conflict. President Taylor subsequently resigned his office and went into hiding in Nigeria in August 2003, until being taken into custody by the Court on 29 March 2006.[19] His trial, which was moved to The Hague, commenced in early 2008.[20] Charles Taylor was found guilty on 26 April 2012 on all 11 counts, when the Trial Chamber delivered its judgement. On 30 May 2012, the Trial Chamber delivered its sentencing judgement, sentencing Taylor to 50 years imprisonment. On 26 September 2013, Taylor's conviction and sentence were upheld by the judgment of the Appeals Chamber.[21]

Pro government forces, most notably the so-called Civilian Defence Forces (CDF) militia, also known as the Kamajors, have also been brought to account for human rights abuses allegedly committed

18   Transcript of Proceedings, 'Opening Statement of the Prosecution', *The Prosecutor v Issa Hassan Sesay, Morris Kallon and Augustine Gbao (Trial)*, (Trial Chamber I, Special Court for Sierra Leone, Case No SCSL-04-15-T, 5 July 2004); Jalloh, C C, 'The Contribution of the Special Court for Sierra Leone to the Development of International Law' (2007) 15 *RADIC* 165, 165–207. See also Schocken, C, 'The Special Court for Sierra Leone: Overview and Recommendations' (2002) 20 *Berkeley Journal of International Law* 436.
19   Brockman, J, 'Liberia: The Case for Changing UN Processes for Humanitarian Interventions' (2004) 22 *Wisconsin International Law Journal* 711, 738–739; Special Court for Sierra Leone, Office of the Prosecutor, 'Chief Prosecutor Announces the Arrival of Charles Taylor at the Special Court' (Press Release, 29 March 2006). Available at: www.rscsl.org/Documents/Press/OTP/prosecutor-032906.pdf.
20   Transcript of Proceedings, 'Evidence of Expert Witness Ian Smillie', *The Prosecutor v Charles Taylor (Trial)*, (Special Court for Sierra Leone, Trial Chamber II, Case No SCSL-03-01-PT, 7 January 2008).
21   The Special Court for Sierra Leone and the Residual Court for Sierra Leone at Freetown and The Hague, *The Prosecutor vs. Charles Gankay Taylor*. Available at: www.rscsl.org/Taylor.html.

during the war. The Trial Chamber delivered its judgement in that case on 2 August 2007, convicting all accused, and the Appeals Chamber confirmed the convictions in its judgement of 28 May 2008.[22]

Another forum for international conflict diamonds prosecutions is the treaty-based permanent ICC. The ICC possesses territorial and nationality jurisdiction for states parties for crimes committed since 2002, and prosecutions in relation to such states parties may be initiated through referral by the particular state party, under Article 13(a), or by the prosecutor acting on its own initiative, under Article 13(c).[23] Along with Sierra Leone, the DRC is perhaps the most important state to have ratified the ICC statute from the perspective of conflict diamonds prosecutions. It should be noted that the court may also exercise jurisdiction over situations in states that are not parties to the statute under limited circumstances. A non-party state may give the court jurisdiction in relation to a 'crime in question', pursuant to Article 12(3), and the court may similarly exercise jurisdiction over a non-party state in the event of a referral from the UNSC, under Article 13(b). The mechanism of a non-party state giving jurisdiction to the court was utilised by Côte d'Ivoire in September 2003, when it requested that the ICC accept jurisdiction over crimes committed on its territory since the events of 19 September 2002 'for an unspecified period of time'. The grant of jurisdiction was confirmed by the Côte d'Ivoire Government in documents dated 14 December 2010 and 3 May 2011. In a decision of Pre-Trial Chamber III of the ICC dated 3 October 2011, the grant of jurisdiction was used as a foundation to approve investigations by the prosecutor into crimes allegedly

---

22 *The Prosecutor v Moinina Fofana and Allieu Kondewa (Trial Judgement)* (Special Court for Sierra Leone, Trial Chamber I, Case No SCSL-04-14-T, 2 August 2007).
23 *Rome Statute of the International Criminal Court*, opened for signature 17 July 1998, 2187 UNTS 90 (entered into force 1 July 2002).

committed by former Côte d'Ivoire President Laurent Gbagbo.[24] His case at the ICC was joined to that of Blé Goudé on 11 March 2015, and the trial commenced on 28 January 2016.[25]

Human rights violations in the diamond-rich Ituri region of the DRC have already attracted the attention of the ICC, leading to indictments and the first arrests by the recently established court.[26] Thomas Lubanga Dyilo, the alleged leader of the military wing of the *Union des Patriotes Congolais*, was arrested on the same day as the issuance of the indictment against him on 17 March 2006, with the assistance of Congolese authorities, French armed forces, and MONUC forces. Dyilo was found guilty on 14 March 2012 of the war crimes of enlisting and conscripting children under the age of 15 years and using them to participate actively in hostilities. He was sentenced on 10 July 2012 to a total of 14 years of imprisonment. His verdict and sentences were confirmed by the ICC Appeals Chamber on 1 December 2014.[27]

The ICC followed up its first arrest with the arrest of Germain Katanga, the alleged military leader of the *Force de Résistance Patriotique en Ituri* (FRPI), on 18 October 2007, also in relation to alleged crimes in the Ituri area. Katanga was found guilty on 7 March 2014 of one count of crime against humanity and 4 counts of war crimes committed on 24 February 2003 during the attack on the village of Bogoro (DRC).

---

24 Ibid; International Criminal Court, Pre-Trial Chamber III, 'Decision Pursuant to Article 15 of the Rome Statute on the Authorisation of an Investigation into the Situation in the Republic of Côte d'Ivoire' (Pre-Trial Chamber III Decision, ICC-02/11, 3 October 2011) paras 10–15, 34–35, 212–213. Available at: www.icc-cpi.int/Pages/record.aspx?docNo=ICC-02/11-14&ln=en; Punyasena, W, 'Conflict Prevention and the International Criminal Court: Deterrence in a Changing World' (2006) 14 *Michigan State Journal of International Law* 39, 64–65; Petrova, P, 'The Implementation and Effectiveness of the Kimberley Process Certification Scheme in the United States' (2006) 40 *International Lawyer* 945, 945; Dawn News, 'Blood Diamond Fears in Ivory Coast Political Duel' (28 December 2010). Available at: www.dawn.com/2010/12/28/blood-diamond-fears-in-ivory-coast-political-duel.html; McClanahan, P, *As Ivory Coast's Gbagbo Holds Firm, 'Blood Diamonds' Flow for Export* (23 January 2011) ReliefWeb. Available at: reliefweb.int/node/381665.
25 The International Criminal Court, *The Prosecutor vs. Laurent Gbagbo and Blé Goudé*, ICC-02/11-01/15. Available at: www.icc-cpi.int/cdi/gbagbo-goude/Pages/default.aspx.
26 Fonseca, A, *Four Million Dead: The Second Congolese War, 1998-2004* (18 April 2004) 49. Available at: www.oocities.org/afonseca/CongoWar.htm.
27 International Criminal Court, 'Case Information Sheet: The Prosecutor v. Thomas Lubanga Dyilo', ICC-01/04-01/06. Available at: www.icc-cpi.int/iccdocs/PIDS/publications/LubangaENG.pdf.

He was sentenced on 23 May 2014 to a total of 12 years imprisonment. The judgement is final as parties have discontinued their appeals. Decisions on possible reparations to victims will be rendered later.[28]

Mathieu Ngudjolo Chui, allegedly the military leader of the *Front des Nationalistes et Intégrationnistes* (FNI), was the third person arrested, on 7 February 2008, in relation to the conflict in this area. Chui was acquitted on 18 December 2012 of three counts of crimes against humanity and seven counts of war crimes. He was released from ICC custody on 21 December 2012. His acquittal was confirmed by the ICC Appeals Chamber on 27 February 2015. [29]

Conflict Diamonds: Use as Context

Prosecutions by the international criminal justice system can take into account the significance of the conflict diamonds trade in several ways. The first is primarily contextual: control over diamond mining and diamond trading becomes a military objective within a conflict. The Sierra Leone cases have highlighted the fact that Kono, Kenema, and Kailahun districts were targeted for combat operations as a result of their being rich areas for diamond mining.[30] The diamond mining areas changed hands several times, between Armed Forces

---

28   International Criminal Court, 'Case Information Sheet: The Prosecutor v. Germain Katanga', ICC-01/04-01/07. Available at: www.icc-cpi.int/iccdocs/PIDS/publications/KatangaEng.pdf.

29   International Criminal Court, 'Case Information Sheet: The Prosecutor v. Mathieu Ngudjolo Chui', ICC-01/04-02/12. Available at: www.icc-cpi.int/iccdocs/PIDS/publications/ChuiEng.pdf. Human rights violations and ongoing conflict have also occurred in the Rwandan-influenced regions of North and South Kivu, although a recent peace deal with General Nkunda, leader of a RCD-Goma splinter group, may herald the way to greater stability. See British Broadcasting Corporation, 'Eastern Congo Peace Deal Signed' (23 January 2008). Available at: www.globalpolicy. org/security/issues/congo/2008/0123gomadeal1.htm. An arrest warrant was also issued for Bosco Ntaganda. 'Warrant of Arrest', *The Prosecutor v Bosco Ntaganda* (International Criminal Court, Case No ICC-01/04-02/06, 7 August 2006); *The Prosecutor v Bosco Ntaganda (Decision on Prosecutor's Application for Warrants of Arrest, Article 58)* (International Criminal Court, Pre-Trial Chamber I, Case No ICC-01/04-02/06, 10 February 2006).

30   *The Prosecutor v Issa Hassan Sesay, Morris Kallon and Augustine Gbao (Trial Judgement)* (Special Court for Sierra Leone, Trial Chamber I, Case No SCSL-04-15-T, 2 March 2009) 3, 8, 10, 184, 186, 188, 239, 251, 348–349 (strategic importance of Kono diamonds fields), 355 (confiscation of diamonds from civilians); *The Prosecutor v Issa Hassan Sesay, Morris Kallon and Augustine Gbao (Appeal Judgement)* (Special Court for Sierra Leone, Appeals Chamber, 29 October 2009) 4, 365–367 (regarding findings on Kailahun as a diamond mining site).

Revolutionary Council (AFRC), RUF, and CDF.[31] Context has also been given through setting out, in both RUF and AFRC indictments, the capture of diamond-rich areas as an objective AFRC/RUF, involving the commission of international crimes as a means to achieving this objective.[32]

When used contextually, the role of conflict diamonds is not articulated as criminal in itself, but rather shows the significance of the trade in exacerbating the conflict and increasing the likelihood that international crimes will be committed. Areas of economic significance may be considered legitimate military targets. International crimes only become involved when, for example, civilians are targeted as part

---

31 Transcript of Proceedings, *The Prosecutor v Charles Taylor (Trial)*, (Special Court for Sierra Leone, Trial Chamber II, Case No SCSL-03-01-PT, 4 June 2007) 59–60; 65; *The Prosecutor v Alex Tamba Brima, Brima Bazzy Kamara and Santigie Borbor Kanu (Trial Judgement)* (Special Court for Sierra Leone, Trial Chamber II, Case No SCSL-04-16-T, 20 June 2007) [336]; Transcript of Proceedings, 'Opening Statement of the Prosecution', *The Prosecutor v Moinina Fofana and Allieu Kondewa (Trial)*, (Special Court for Sierra Leone, Trial Chamber I, Case No SCSL-04-14-T, 3 June 2003) 10, 11, 20, 28; *The Prosecutor v Moinina Fofana and Allieu Kondewa (Trial Judgement)* (Special Court for Sierra Leone, Trial Chamber I, Case No SCSL-04-14-T, 2 August 2007) [374], [381], [384], [389]–[405]; *The Prosecutor v Issa Hassan Sesay, Morris Kallon and Augustine Gbao (Trial Judgement)* (Special Court for Sierra Leone, Trial Chamber I, Case No SCSL 04 15 T, 2 March 2009) 612, 614 (concerns the rift between AFRC and RUF which occurred in April 1998).
32 'Corrected Amended Consolidated Indictment', *The Prosecutor v Issa Hassan Sesa, Morris Kallon and Augustine Gbao* (Special Court for Sierra Leone, Case No SCSL-2004-15-PT, 2 August 2006) [36]–[39], [71] cited in *The Prosecutor v Issa Hassan Sesay, Morris Kallon and Augustine Gbao (Appeal Judgement)* (Special Court for Sierra Leone, Appeals Chamber, 29 October 2009) 6, 28; see also the identical wording in 'Further Amended Consolidated Indictment', *The Prosecutor v Alex Tamba Brima, Brima Bazzy Kamara and Santigie Borbor Kanu* (Special Court for Sierra Leone, Case No SCSL-2004-16-PT, 18 February 2005) [33] cited in *The Prosecutor v Alex Tamba Brima, Brima Bazzy Kamara and Santigie Borbor Kanu (Appeal Judgement)* (Special Court for Sierra Leone, Appeals Chamber, Case No SCSL-2004-16-A, 22 February 2008) 81. The Prosecution in the RUF Indictment sought to modify the purpose to be the 'pillage the resources in Sierra Leone, particular diamonds [sic], and to control forcibly the population and territory of Sierra Leone'. However, this modification was rejected by the Trial Chamber in its judgment. *The Prosecutor v Issa Hassan Sesay, Morris Kallon and Augustine Gbao (Trial Judgement)* (Special Court for Sierra Leone, Trial Chamber I, Case No SCSL-04-15-T, 2 March 2009) 115, 118, 127–128. Nevertheless, the chamber held that it did not materially prejudice the defence case. It is also apparent that the confirmation of the original formulation by the chamber did not have an adverse effect on the prosecution. The Trial Chambers' approach was upheld on appeal. *The Prosecutor v Issa Hassan Sesay, Morris Kallon and Augustine Gbao (Appeal Judgement)* (Special Court for Sierra Leone, Appeals Chamber, 29 October 2009) 35–38. The Appeals Chamber, furthermore, confirmed the Trial Chamber's finding that a common purpose can be constituted by a non-criminal objective (i.e. diamond mining), where the intended means to achieve that objective are criminal (i.e. enslavement, terror, etc.): *The Prosecutor v Issa Hassan Sesay, Morris Kallon and Augustine Gbao (Appeal Judgement)* (Special Court for Sierra Leone, Appeals Chamber, 29 October 2009) 107, see also 120–121,127–128, 130–135.

of the military operation, when child soldiers are employed as part of the operation, or when slave labour is used to mine diamonds in the area.

The Charles Taylor case also used the trade in conflict diamonds to provide a context for the commission of international crimes. The indictment focused on six fields controlled at different times by the RUF. In Kono, there were diamond fields in Koidu, Tombudu, and Yengema. In Kenema, there were the Tongo Field, including the so-called 'Cyborg Pit'.[33] In its opening statement against Taylor, the prosecution argued that Taylor conducted his diamond transactions through Eddie Kanneh, a Sierra Leonean and former SLA officer who joined the RUF in 1998.[34] The statement alleged that in 2000 there were regular shipments of arms, in exchange for diamonds, from Taylor to the RUF in Sierra Leone, and Taylor's men visited the RUF-held territories and reported to Taylor on economic and military developments.[35] Similarly, the CDF case describes military operations to wrest control of the Tongo diamond fields from the RUF and AFRC as the context for crimes that were committed by the CDF or Kamajors at those locations.[36]

The conflict diamonds cases before the ICC also refer to the contextual role of diamonds as military targets in the war in the DRC. The charging document for the Katanga and Ngudjolo cases refers in the background section on the 'Region of Ituri' to the significance of natural resources, including diamonds, in exacerbating the conflict in the region. It states that the desire to control Ituri's natural resources has been integral in promoting conflict in the region. The charging document states that

33   Transcript of Proceedings, *The Prosecutor v Charles Taylor (Trial)*, (Special Court for Sierra Leone, Trial Chamber II, Case No SCSL-03-01-PT, 4 June 2007) 33. See also Transcript of Proceedings, 'Opening Statement of the Prosecution', *The Prosecutor v Issa Hassan Sesay, Morris Kallon and Augustine Gbao (Trial)*, (Trial Chamber I, Special Court for Sierra Leone, Case No SCSL-04-15-T, 5 July 2004) 48; *The Prosecutor v Issa Hassan Sesay, Morris Kallon and Augustine Gbao (Trial Judgement)* (Special Court for Sierra Leone, Trial Chamber I, Case No SCSL-04-15-T, 2 March 2009) 323–324, 439–440.
34   Transcript of Proceedings, *The Prosecutor v Charles Taylor (Trial)*, (Special Court for Sierra Leone, Trial Chamber II, Case No SCSL-03-01-PT, 4 June 2007) 40, 42, 45, 49–51.
35   Ibid 51–53.
36   *The Prosecutor v Moinina Fofana and Allieu Kondewa (Trial Judgement)* (Special Court for Sierra Leone, Trial Chamber I, Case No SCSL-04-14-T, 2 August 2007) [375]; *The Prosecutor v Moinina Fofana and Allieu Kondewa (Appeal Judgement)* (Special Court for Sierra Leone, Appeals Chamber, Case No SCSL-04-14-A, 28 May 2008) 16, 80; *The Prosecutor v Alex Tamba Brima, Brima Bazzy Kamara and Santigie Borbor Kanu (Trial Judgement)* (Special Court for Sierra Leone, Trial Chamber II, Case No SCSL-04-16-T, 20 June 2007) [327]–[332].

the natural wealth in Ituri includes gold, diamonds, Colombo tantalite (coltan), timber and oil. Mongbwalu, located north-west of Bunia in Djugu territory, is known as an important gold mine.[37] In the context of the Katanga and Ngudjolo cases, the connection between the two local military forces to Ugandan and DRC governments are stated to have been political and military in nature although it is known from UN reports that the relationship also involved trading in diamonds and other natural resources.[38]

The Lubanga arrest warrant does not directly refer to conflict diamonds other than in a contextual sense. In terms of substantive crimes, Lubanga is charged with the enlistment, conscription, and use of child soldiers in the conflict in Ituri in the north-east of the DRC. However, the diamond trade is referred to in the prosecutor's opening statement, as part of the factual context in which the child soldiers were recruited and deployed.[39]

Conflict Diamonds: Substantive Crimes

One way in which conflict diamonds situations may accrue international criminal liability is when the mining itself involves criminal conduct. Key examples of this are when the RUF and the AFRC conducted mining operations using abducted civilians as slaves,

---

37 'Document Containing the Charges Pursuant to Article 61(3)(a) of the Statute', *The Prosecutor v Germain Katanga and Mathieu Ngudjolo Chui* (International Criminal Court, Case No ICC-01/04-01/07, 21 April 2008) [1], [16], [25]–[35], [72]–[89], 31–34; International Criminal Court, 'Case Information Sheet: The Prosecutor v. Germain Katanga', ICC-01/04-01/07. Available at: www. icc-cpi.int/iccdocs/PIDS/publications/KatangaEng.pdf. See also 'Warrant of Arrest', *Ntaganda* (International Criminal Court, Case No ICC-01/04-02/06, 7 August 2006) 4, 5; (International Criminal Court, Pre-Trial Chamber I, Case No ICC-01/04-02/06, 10 February 2006) 12.

38 'Document Containing the Charges Pursuant to Article 61(3)(a) of the Statute', *Chui* (International Criminal Court, Case No ICC-01/04-01/07, 21 April 2008) [1], [16], [25]–[35], [72]-[89], 31–34; Final Report Panel of Experts on the Illegal Exploitation of Natural Resources and Other Forms of Wealth of the Democratic Republic of the Congo, UN Doc S/2002/1146 (16 October 2002).

39 'Warrant of Arrest', *The Prosecutor v Thomas Lubanga Dyilo* (International Criminal Court, Case No ICC-01/04-01/06, 10 February 2006) 1–5; International Criminal Court, 'Case Information Sheet: The Prosecutor v. Thomas Lubanga Dyilo', ICC-01/04-01/06. Available at: www.icc-cpi. int/iccdocs/PIDS/publications/LubangaENG.pdf ; Transcript of Proceedings, 'Prosecutor's Opening Statement', *The Prosecutor v Thomas Lubanga Dyilo (Trial)*, (International Criminal Court, Trial Chamber I, Case No ICC-01/04-01/06, 26 January 2009) 15.

with child soldiers operating as enforcers of the slave labour system.[40] Small mining communities were called zoo bushes.[41] Conditions for the miners were harsh, with there often being little or no food available for their sustenance, and miners were sometimes forced to wear only their underwear in an effort by the RUF to exert authority over them.[42] Civilians were killed or beaten at mines such as the Cyborg Pit, some because they were suspected of stealing diamonds, and others because their deaths created a climate of terror to deter escape.[43]

A significant, if controversial, finding of the RUF Trial Chamber was that it determined that the forced labour that occurred at the Kenema district mining sites, especially the Cyborg Pit, constituted the war crime of terror whereas, by contrast, in the Kono district it was determined that, although enslavement had been made out, terror did not occur. The Trial Chamber drew a distinction between the severity of the enslavement in Kenema and Kono: whereas the former

---

40   'Corrected Amended Consolidated Indictment', *The Prosecutor v Issa Hassan Sesa* (Special Court for Sierra Leone, Case No SCSL-2004-15-PT, 2 August 2006) 8–22, especially [70]–[71]; *The Prosecutor v Issa Hassan Sesay, Morris Kallon and Augustine Gbao (Trial Judgement)* (Special Court for Sierra Leone, Trial Chamber I, Case No SCSL-04-15-T, 2 March 2009) 298, 334–338 (describing the use of the 'Small Boys Unit' to kill and terrorise civilians used to forcibly mine at the Cyborg Pit mine and the Tongo Field mine. It also discusses forced mining in Kono district), 340–341, 343–344 (use of forced mining labour as 'enslavement'), 346–347 (discusses killings at pit as 'terror'), 348 (killings at Cyborg Pit not considered 'collective punishment'), 375–382, 398–399, 423, 429–430, 443 (forced mining as 'enslavement' in Kailahun District), 497–498 (use of child soldiers in Tongo Field, Kenema, to guard diamond mining operations: they committed most of the documented killings there), 511–514, 584, 635; Transcript of Proceedings, 'Opening Statement of the Prosecution', *The Prosecutor v Alex Tamba Brima, Brima Bazzy Kamara and Santigie Borbor Kanu (Trial)*, (Special Court for Sierra Leone, Trial Chamber I, Case No SCSL-2004-16-T, 7 March 2005) 24, 25, 30, 35, 36; Transcript of Proceedings, *The Prosecutor v Charles Taylor (Trial)*, (Special Court for Sierra Leone, Trial Chamber II, Case No SCSL-03-01-PT, 4 June 2007) 54, 65–67; 'Further Amended Consolidated Indictment', *The Prosecutor v Alex Tamba Brima* (Special Court for Sierra Leone, Case No SCSL-2004-16-PT, 18 February 2005) 67–68.

41   *The Prosecutor v Issa Hassan Sesay, Morris Kallon and Augustine Gbao (Trial Judgement)* (Special Court for Sierra Leone, Trial Chamber I, Case No SCSL-04-15-T, 2 March 2009) 423.

42   Transcript of Proceedings, *The Prosecutor v Charles Taylor (Trial)*, (Special Court for Sierra Leone, Trial Chamber II, Case No SCSL-03-01-PT, 4 June 2007) 64; *The Prosecutor v Issa Hassan Sesay, Morris Kallon and Augustine Gbao (Trial Judgement)* (Special Court for Sierra Leone, Trial Chamber I, Case No SCSL-04-15-T, 2 March 2009) 334–338, 340–341, 343–344, 346–348, 375–382, 398–399.

43   Transcript of Proceedings, *The Prosecutor v Charles Taylor (Trial)*, (Special Court for Sierra Leone, Trial Chamber II, Case No SCSL-03-01-PT, 4 June 2007) 54, 62, 65-67; 'Further Amended Consolidated Indictment', *The Prosecutor v Alex Tamba Brima* (Special Court for Sierra Leone, Case No SCSL-2004-16-PT, 18 February 2005) [67]–[68]; *The Prosecutor v Issa Hassan Sesay, Morris Kallon and Augustine Gbao (Trial Judgement)* (Special Court for Sierra Leone, Trial Chamber I, Case No SCSL-04-15-T, 2 March 2009) 334–338, 340–341, 343–344, 346–348, 356–357, 372, 375–382, 398–399.

was characterised by severe mistreatment and killings, the latter was considered less severe. These factual findings played in the Trial Chamber's mind when it was called upon to make its legal finding as to whether the actions of the AFRC/RUF showed a specific intent to cause terror. Due to the difference in severity, the chamber found that there was no specific intent to cause terror in Kono, terror being a side-effect of the enslavement, although it was present in Kenema.[44] This distinction, however, appears a little artificial. It is recalled that what must be proved is the 'specific intent to cause terror'. When it is clear that the consequence of large-scale abduction and enslavement is the creation of terror in a civilian population, it is difficult to argue that terror was not also intended when the campaign of enslavement was implemented.

This discussion on the nature of the specific intent to cause terror is similar to the legal argument that if there is a clear objective in mind underlying a crime such as the obtaining of a military objective, then it is not possible to determine that the action was carried out with the specific intent to cause terror. This reasoning, however, was overruled by the Appeals Chamber that concluded the specific intent to cause terror can exist side-by-side with other objectives.[45]

The mining of diamonds may also be considered a crime in itself if it is categorised as the plunder of the natural resources of the state concerned. Jurisprudence on the war crime of plunder and pillage articulates that it is not simply private property that might be open to being plundered, but also public or state-controlled property. Pursuant to domestic legislation, natural resources are typically considered to be owned by the state rather than being private property, subject to licensing out to individuals for commercial exploitation purposes.[46] Therefore, the ransacking of a diamond mine or unauthorised

44 *The Prosecutor v Issa Hassan Sesay, Morris Kallon and Augustine Gbao (Trial Judgement)* (Special Court for Sierra Leone, Trial Chamber I, Case No SCSL-04-15-T, 2 March 2009) 407.
45 Ibid [1348]; *The Prosecutor v Issa Hassan Sesay, Morris Kallon and Augustine Gbao (Appeal Judgement)* (Special Court for Sierra Leone, Appeals Chamber, 29 October 2009) 240–242, 318–319.
46 *Loi Portant Code Minier 2002* [Law Relating to the Mining Code] (The Democratic Republic of Congo) Law No 007/2002, 11 July 2002; *Da Lei dos Diamantes 1994* [The Diamond Law] (Angola) Law No 1/92, 7 October 1994; *Code Minier 1995* [Mining Code] (Côte d'Ivoire) Law No 95-553, 17 July 1995; *Mines and Minerals Act 1994* (Sierra).

exploitation of an alluvial diamond field might be considered an act of plunder against the state as well as or instead of the theft of private property.[47]

An opportunity was missed to test the parameters of the war crime of pillage in the context of the work of the Sierra Leone Special Court. Unfortunately, count 14 of the indictment charged pillage of 'civilian property' in the Kono district and did not refer to the diamond resources of Sierra Leone. Even though the prosecution made submissions on this charge in its prosecution final trial brief, the chamber refused to consider them due to the lack of particularisation in the indictment.[48] At the trial level, the chamber found that this type of pillaging did not constitute an act of terrorism as it lacked, in their opinion, the specific intent to spread terror. The chamber found that, as the name of the military operation, 'Operation Pay Yourself', suggests, the AFRC/RUF rebels appropriated civilian property for their personal gain.[49] This finding, however, is open to the criticism that, although one of the objectives of Operation Pay Yourself was personal gain, it appears clear that terror is a consequence of such behaviour. Therefore, it is possible to consider that a charge of terrorism should have also been made out in relation to this behaviour.

Interestingly, the chamber had to determine whether armed conduct by child soldiers against civilians at diamond mines could be considered to be active participation in hostilities, as is required for the crime of using child soldiers. Naturally, such attacks contravene the international law prohibition on the targeting of civilians, and so a question arises whether such attacks can be included in the concept of active hostilities. Significantly, the chamber did not say that, in doing so, the civilians were legitimate military targets, but simply that there were strategic outcomes in military terms which accrued to the RUF through deploying the child soldiers in this manner.[50]

---

47 *The Prosecutor v Issa Hassan Sesay, Morris Kallon and Augustine Gbao (Trial Judgement)* (Special Court for Sierra Leone, Trial Chamber I, Case No SCSL-04-15-T, 2 March 2009) 399, discussion of the crime of 'pillage'. The chamber did not consider pillage in the context of looting state property. In an interesting discussion, the chamber noted the widespread nature of the looting to support it as being a 'serious violation', therefore meeting the test set out in the jurisdiction of the Special Court (as distinct from international law).
48 Ibid 400–401.
49 Ibid 408.
50 Ibid 511–513.

Conflict Diamonds: Indirect Liability

Another way in which conflict diamonds prosecutions can be employed is in establishing individual criminal responsibility for other international crimes. For example, Charles Taylor is alleged to have assisted the RUF by accepting diamonds in exchange for providing weapons and ammunition.[51] Diamond transactions that take place in the knowledge that the transaction will assist the RUF to continue to commit international crimes connect the individual indirectly to the crimes ultimately committed by the RUF, such as the unlawful killing of civilians, causing bodily harm to civilians, the use of child soldiers, or sexual offences.[52] A supplementary question is how far down the line it might be possible to prosecute an individual for indirectly assisting the commission of international crimes in this way.

Both the ICC statute and the statute of the Sierra Leone Special Court support the application of the recently developed doctrine of joint criminal enterprise. This doctrine has arisen from the recent jurisprudence of the Yugoslav Tribunal and the Rwanda Tribunal and is a doctrine of individual criminal responsibility that would be of great utility in pursuing a prosecution in relation to the conflict diamonds problem.[53] A person may be guilty by supporting a criminal project, for example through the provision of the economic resources needed to sustain the project, even if the person does not directly carry out the crime.

The international criminal law requirements for joint criminal enterprise liability were recently set out in a case before the International Criminal Tribunal for the former Yugoslavia Appeals Chamber. Regardless of the category at issue, or of the charge under consideration, a conviction requires a finding that the accused

---

51 Transcript of Proceedings, *The Prosecutor v Charles Taylor (Trial)*, (Special Court for Sierra Leone, Trial Chamber II, Case No SCSL-03-01-PT, 4 June 2007) 30–31, 41, 74–78. See also *The Prosecutor v Issa Hassan Sesay, Morris Kallon and Augustine Gbao (Trial Judgement)* (Special Court for Sierra Leone, Trial Chamber I, Case No SCSL-04-15-T, 2 March 2009) 259–260, 264, 266, 595–596, regarding the RUF/AFRC trade in diamonds.

52 Transcript of Proceedings, 'Opening Statement of the Prosecution', *The Prosecutor v Issa Hassan Sesay, Morris Kallon and Augustine Gbao (Trial)*, (Trial Chamber I, Special Court for Sierra Leone, Case No SCSL-04-15-T, 5 July 2004) 20–21, 25–26, 39.

53 *Rome Statute of the International Criminal Court*, opened for signature 17 July 1998, 2187 UNTS 90 (entered into force 1 July 2002) art 25(3)(d); *The Prosecutor v Issa Hassan Sesay, Morris Kallon and Augustine Gbao (Trial Judgement)* (Special Court for Sierra Leone, Trial Chamber I, Case No SCSL-04-15-T, 2 March 2009) [257], [1977]–[1985].

participated in a joint criminal enterprise. There are three requirements for such a finding. First is a plurality of persons. Second is the existence of a common purpose (or plan) that amounts to or involves the commission of a crime provided for in the statute. Third is the participation of the accused in this common purpose, characterised as the accused making a 'significant contribution' to the common purpose. The *mens rea* required for a finding of guilt differs according to the category of joint criminal enterprise liability under consideration. Where convictions under the first category of joint criminal enterprise are concerned, the accused must both intend the commission of the crime and intend to participate in a common plan aimed at its commission.[54]

The indictments for both the RUF and AFRC cases refer to the role of the diamond trade as being a goal of the joint criminal enterprise in Sierra Leone:

> The RUF ... shared a common plan ... which was to take any actions necessary to gain and exercise political power and control over the territory of Sierra Leone, in particular the diamond mining areas. The natural resources of Sierra Leone, in particular the diamonds, were to be provided to persons outside Sierra Leone in return for assistance in carrying out the joint criminal enterprise.[55]

---

54   *The Prosecutor v Brdjanin (Appeal Judgement) (International Criminal Tribunal for the Former Yugoslavia*, Appeals Chamber, Case No IT-99-36-A, 3 April 2007) [364]–[365]. See also [227]–[228].

55   'Corrected Amended Consolidated Indictment', *The Prosecutor v Issa Hassan Sesa* (Special Court for Sierra Leone, Case No SCSL-2004-15-PT, 2 August 2006) [36]–[39], [71] cited in *The Prosecutor v Issa Hassan Sesay, Morris Kallon and Augustine Gbao (Appeal Judgement)* (Special Court for Sierra Leone, Appeals Chamber, 29 October 2009) 6, 28. See also the identical wording in 'Further Amended Consolidated Indictment', *The Prosecutor v Alex Tamba Brima* (Special Court for Sierra Leone, Case No SCSL-2004-16-PT, 18 February 2005) [33] cited in *The Prosecutor v Alex Tamba Brima, Brima Bazzy Kamara and Santigie Borbor Kanu (Appeal Judgement)* (Special Court for Sierra Leone, Appeals Chamber, Case No SCSL-2004-16-A, 22 February 2008) [81]. The prosecution in the RUF indictment sought to modify the purpose to be the 'pillage the resources in Sierra Leone, particular diamonds [sic], and to control forcibly the population and territory of Sierra Leone'. However, this modification was rejected by the Trial Chamber in its judgement. *The Prosecutor v Issa Hassan Sesay, Morris Kallon and Augustine Gbao (Trial Judgement)* (Special Court for Sierra Leone, Trial Chamber I, Case No SCSL-04-15-T, 2 March 2009) 115, 118, 127–128. Nevertheless, the chamber held that it did not materially prejudice the defence case. It is also apparent that the confirmation of the original formulation by the chamber did not have an adverse effect on the prosecution. The Trial Chambers' approach was upheld on appeal. *The Prosecutor v Issa Hassan Sesay, Morris Kallon and Augustine Gbao (Appeal Judgement)* (Special Court for Sierra Leone, Appeals Chamber, 29 October 2009) 35–38. The Appeals Chamber, furthermore, confirmed the Trial Chamber's finding that a common purpose can be constituted by a non-criminal objective (i.e. diamond mining), where the intended means to achieve that objective are criminal (i.e. enslavement, terror etc.): *The Prosecutor v Issa Hassan Sesay, Morris Kallon and Augustine Gbao (Appeal Judgement)* (Special Court for Sierra Leone, Appeals Chamber, 29 October 2009) 107, see also 120–121, 127–128, 130–135.

It is interesting to note that, in the AFRC case, the reference to conflict diamonds as part of the joint criminal enterprise was challenged on the grounds of being irrelevant to the crimes pleaded in the document. It was argued that the objectives of the common plan must themselves be crimes. This argument was upheld by the Trial Chamber, but the Appeals Chamber overturned it, arguing instead that, while the common plan of a joint criminal enterprise must involve the commission of international crimes, there may be other elements involved in the joint criminal enterprise that are not intrinsically criminal.[56] By contrast, the CDF indictment does not refer to diamonds, although it mentions the objective of gaining and exercising control over the territory of Sierra Leone.[57]

April 1998 was a central moment in the conflict as it was the time that the coalition between the RUF and the AFRC ended. In the light of the split between the two factions, the Trial Chamber considered that this was the time that the joint criminal enterprise between the two factions ceased to exist. As a result, different modes of individual liability needed to be pleaded in relation to crimes committed after this date.[58]

In the RUF case, the chamber made findings not only that the control of the diamond mining in Sierra Leone was a part of the common purpose of the joint criminal enterprise, but also that management of the diamond mining by the accused Sesay represented his significant contribution to that joint criminal enterprise. The chamber noted that Sesay planned the enslavement of civilian miners and the use of child soldiers to guard mining sites and force the miners to work at Tongo Field. It is notable that some of these indicia of 'significant contribution' relate to direct contribution to the joint criminal enterprise rather than this mode of liability representing solely an indirect mode of responsibility.

---

56 'Further Amended Consolidated Indictment', *The Prosecutor v Alex Tamba Brima* (Special Court for Sierra Leone, Case No SCSL-2004-16-PT, 18 February 2005) [32]–[33], [74]–[76]; *The Prosecutor v Alex Tamba Brima, Brima Bazzy Kamara and Santigie Borbor Kanu (Appeal Judgement)* (Special Court for Sierra Leone, Appeals Chamber, Case No SCSL-2004-16-A, 22 February 2008) [188].

57 'Indictment', *The Prosecutor v Samuel Hinga Norman, Moinina Fofana and Allieu Kondewa* (Special Court for Sierra Leone, Case No SCSL-03-14-I, 5 February 2004) [19].

58 *The Prosecutor v Issa Hassan Sesay, Morris Kallon and Augustine Gbao (Trial Judgement)* (Special Court for Sierra Leone, Trial Chamber I, Case No SCSL-04-15-T, 2 March 2009) 614–618.

In making this finding, the chamber noted that forced mining in Tongo Field provided an important source of revenue for the junta regime and that this topic was discussed in the AFRC Supreme Council meetings when Sesay was present. The sheer scale of the enslavement in Kenema district demonstrated that the forced mining was a planned and a systematic policy of the regime devised at the highest level. The chamber inferred from Sesay's membership of the Supreme Council that he was involved in the planning and organisation of the forced mining in Kenema. The chamber also found that he, along with Samuel Bockarie, received diamonds at the AFRC Secretariat originating from Tongo Field. In addition, Sesay was personally engaged in mining for his personal benefit in Tongo Field.[59]

The chamber also found that Sesay was involved in mining activities in Kono district and made a significant contribution to the joint criminal enterprise activities there. He visited the mines to collect diamonds, signed off on the mining log books and transported diamonds to Bockarie and also took them to Liberia. The chamber held that Sesay, therefore, participated in the forced labour in diamond mines in Kono district between 14 February and May 1998 in order to further the common purpose.[60]

The chamber also found that Kallon made a significant contribution to the joint criminal enterprise through assisting the diamond mining operations. Firstly, it found that, as a member of the AFRC Supreme Council, Kallon was involved in decision-making processes that included the orchestration of the widespread forced labour in the Kenema district. The chamber also found that Kallon used his bodyguards to force civilians to mine diamonds at Tongo Field, a practice that was prevalent among senior RUF and AFRC commanders. The chamber also found that on two occasions, Kallon was present at the mining pits in Tongo Field when Small Boys Units and other rebels shot into the pits, killing unarmed enslaved civilian miners.[61]

---

59   *The Prosecutor v Issa Hassan Sesay, Morris Kallon and Augustine Gbao (Trial Judgement)* (Special Court for Sierra Leone, Trial Chamber I, Case No SCSL-04-15-T, 2 March 2009) 588–589, 604–605, 611.

60   Ibid 618–619.

61   Ibid 590–591, 604–605, 611. See also *The Prosecutor v Issa Hassan Sesay, Morris Kallon and Augustine Gbao (Appeal Judgement)* (Special Court for Sierra Leone, Appeals Chamber, 29 October 2009) 279, 452–453, 458–459.

In relation to the Kono district, the chamber also found that Kallon made a significant contribution to the joint criminal enterprise through his activities in engaging his bodyguards to supervise on his behalf the private mining by enslaved civilians. Kallon also visited mining sites in Kono district during February/March 1998. He also received regular communications about the activities of the joint forces in Kono. As a result, the chamber found that Kallon actively participated in the joint criminal enterprise in Kono.[62]

In relation to the period from December 1998 to January 2000, the chamber found that Bockarie appointed M S Kennedy as the Overall Mining Commander in Kono district. In 2000, it was Sesay who appointed Kennedy's replacement. The overall mining commander reported to Sesay. Throughout 1999 and 2000, Sesay visited Kono district and collected diamonds. Sesay maintained a house in Koidu Town where he received mining commanders for this purpose. He also visited the mines and ordered that civilians be captured from other districts. He arranged for transportation of the captured civilians to the mines. The chamber found that the nature and magnitude of the forced mining in Kono district required extensive planning on an ongoing basis. It was provided by the detailed administrative and archiving records maintained to compute the size, grade, origin, and value of the diamonds found. The mining system in Kono district was designed and supervised by Sesay who operated at the highest levels. His conduct was a significant contributory factor to the perpetration of enslavement, and he intended the commission of the crimes.[63] On appeal, these findings were challenged on the basis that the chamber relied on planning as the mode of individual criminal liability for Sesay. The Appeal Chamber found that the Trial Chamber did rely on planning, but that the legal test for planning was correctly identified as a substantial contribution rather than a significant contribution.[64]

62   *The Prosecutor v Issa Hassan Sesay, Morris Kallon and Augustine Gbao (Trial Judgement)* (Special Court for Sierra Leone, Trial Chamber I, Case No SCSL-04-15-T, 2 March 2009) 621. See also *RUF Appeal* (Special Court for Sierra Leone, Appeals Chamber, 29 October 2009) 279, 452–453, 458–459.

63   *The Prosecutor v Issa Hassan Sesay, Morris Kallon and Augustine Gbao (Trial Judgement)* (Special Court for Sierra Leone, Trial Chamber I, Case No SCSL-04-15-T, 2 March 2009) 624. See also general conclusion regarding the JCE at 639.

64   *The Prosecutor v Issa Hassan Sesay, Morris Kallon and Augustine Gbao (Appeal Judgement)* (Special Court for Sierra Leone, Appeals Chamber, 29 October 2009) 245, 248, 350–251.

An alternate mode of liability is aiding and abetting, which requires that the acts in question must be specifically directed to assist the crime and must make a substantial contribution to the commission of the crime. This mode of liability was used in a civil action taken pursuant to the US Alien Torts Claims Act, in relation to the alleged involvement of a multinational corporation in human rights abuses by government forces providing security services.[65]

Neither the ICC statute nor the Special Court statute provide for the prosecution of corporations as legal persons but allow for the prosecution of individuals who may be the directors of such corporations. This is arguably a weakness in the utilisation of criminal sanctions to stop the conflict diamonds trade, although the possibility of individual prosecution is arguably a stronger deterrent for the corporate leadership. A case scenario considered in the literature was a hypothetical prosecution of De Beers for their involvement with Angolan conflict diamonds under the US Alien Torts Law. The case study found that this would be unlikely to succeed, however, based on a number of issues specific to US domestic law.[66] It is interesting to note that, in relation to the conflict situation in the Congo, the UNSC expert report highlighted poor behaviour not only by individuals but corporations as well. There remains an option for the directors of such corporations, if not the corporations themselves, to face criminal charges before the ICC.[67]

Concluding Remarks

This chapter has considered the organisations above and beyond the Kimberley Process that make a significant contribution to the conflict diamonds governance system. While the United Nations General Assembly provides important political support and lends legitimacy to the KP, it is perhaps the UNSC that is more central to its operation. The UNSC was alert to the issue of conflict diamonds before the KP was established and, in some respects, acted as the midwife to the KP

---

65   *Doe I v Unocal Corp*, 248 F 3d 915, (9th Cir, 2001).

66   Saunders, L, 'Note: Rich and Rare are the Gems they War: Holding De Beers Accountable for Trading Conflict Diamonds' (2001) 24 *Fordham International Law Journal* 1402.

67   Final Report Panel of Experts on the Illegal Exploitation of Natural Resources and Other Forms of Wealth of the Democratic Republic of the Congo, UN Doc S/2002/1146 (16 October 2002).

through its resolutions. The UNSC has lent significant resources to the task of monitoring through its expert committees on the various African conflict situations, as well as enforcement, in particular, through the imposition of diamond trading embargoes. The ultimate ratchet in the international pyramid of sanctions is a prosecution before an international criminal tribunal. This ratchet has been applied in the four cases heard by the Special Court for Sierra Leone, as well as preliminary proceedings in a number of cases before the ICC. Conflict diamonds have been referred to so as to provide the context in which crimes were committed (diamond areas as military targets), as an integral part of the direct commission of crimes (such as the use of child soldiers to enforce diamond mining) and so as to prove indirect liability for the commission of crimes by others (diamond sales to purchase weapons used to commit crimes).

The UNSC and the ICC can be viewed as part of a single conflict diamonds governance system that reinforces the activities of the Kimberley Process. A conflict diamond is conceptualised with reference to the connection between the mining of the stone and grave human rights abuses amounting to international crimes. As such, conflict diamonds inherently attract the jurisdiction of international criminal tribunals. This represents a big stick that, according to pyramid theory, reinforces ability of the Kimberley Process to operate through less coercive means such as negotiation or informal naming and shaming.

6

Raging Bulls and Flyswatters: The Networked Pyramid Model

A fly should not be hit with a sledgehammer, nor a raging bull with a flyswatter.

I Ayres and J Braithwaite, regulatory theorists[1]

Chapter Overview

This chapter sets out the theoretical framework that is employed in later chapters to analyse the effectiveness of the conflict diamonds governance system, thereby responding to the second of two main research questions being considered in this book. The chapter begins by discussing the value of using a regulatory approach in this type of context as opposed to, for example, a strictly legal analysis. Finally, a number of sophisticated regulatory models are explored in more detail, namely the network model, the pyramid model, and an approach that combines the two, the networked pyramid hybrid model.

---

1   Ayres, I and J Braithwaite, *Responsive Regulation: Transcending the Deregulation Debate* (Oxford University Press, 1992) 49.

Why Use a Regulatory Approach?

There are a number of different ways in which a governance system, such as the conflict diamonds governance system, can be analysed. Naturally, the starting point for those with legal training is to identify sources of law, such as treaties, legislation and jurisprudence, from which rights and obligations might be identified. By contrast, a regulatory approach provides a significantly different perspective. Julia Black's definition of regulation is illuminating in this regard:

> Regulation is the sustained and focussed attempt to alter the behaviour of others according to defined standards or purposes, with the intention of producing a broadly identified outcome or outcomes, which may involve mechanisms of standard-setting, information-gathering and behaviour modification.[2]

Consideration of this definition is of assistance in articulating the value of using a regulatory lens to analyse a legal system, whether it be national or international in nature. Of central interest is that regulation has defined standards or purposes that it seeks to achieve. Whereas a lawyer may be content with considering the question, 'what is the law?', a regulatory approach will critically analyse the current law in terms of whether or not it is achieving a particular purpose and, if it is not achieving that purpose, suggest ways in which it might be improved. Regulation extends beyond law itself, to encompass a variety of means by which its purpose might be achieved. For example, a lawyer might question whether an industry code of conduct is legally binding. However, if it has the effect of achieving a particular purpose in that industry, then it may still qualify as effective regulation.[3]

A related consideration is that legal scholars are concerned, to a significant degree, in maintaining the internal consistency and integrity of the rules system. This objective seems to be of comparatively less importance from the regulatory perspective, which is less concerned with whether law is correct in seeing itself

---

2   Black, J, 'Critical Reflections on Regulation', *Australian Journal of Legal Philosophy*, vol. 27, 2002, 20.

3   Parker, C et al. (eds), *Regulating Law* (Oxford University Press, 2004) 1–3.

as characterised by unity, coherence, or particular modes of reasoning. Rather, it is the outcomes of the legal system or regulatory system that are paramount, such as social justice, or economic efficiency.[4]

Regulatory theory not only seeks to analyse a system in terms of its ability to achieve a particular purpose, but also involves a generally agreed set of functional criteria as to whether that system will be successful. In developing these criteria, regulatory theory has borrowed from the study of artificial intelligence, or cybernetics. These criteria can be listed as mechanisms of standard-setting, information gathering (or monitoring) and behaviour modification. These criteria, it is argued, provide for the means of control whereby a system, whether artificial or natural, is kept within a preferred subset of all possible states. In the absence of any one of these elements, there is not control in a cybernetic sense.[5]

Further to the discussion in earlier chapters of this book, the conflict diamonds governance system is defined to include those persons, corporations, and organisations involved in addressing the conflict diamonds issue through their regulatory behaviour. The main players in the system are national governments, the Kimberley Process Certification Scheme (KP) (including the non-governmental organisations (NGOs), corporations and national governments therein represented), the United Nations Security Council (UNSC), and the International Criminal Court (ICC). In considering the utility of applying a regulatory approach to a system such as the conflict diamonds governance system, the regulatory approach will not simply provide insight into the rights and obligations that comprise that system, but will also assess the system in terms of whether it is achieving a particular purpose or set of purposes. In relation to the conflict diamonds governance system, the central issue is whether the system is actually preventing the illegal diamonds trade from providing financial support to human rights abusers.

---

4   Black, J, 'Critical Reflections on Regulation', *Australian Journal of Legal Philosophy*, vol. 27, 2002, 22–26; Parker, C et al. (eds), *Regulating Law* (Oxford University Press, 2004) 3–4.
5   Black, J, 'Critical Reflections on Regulation', *Australian Journal of Legal Philosophy*, vol. 27, 2002, 20.

A regulatory approach also contemplates approaches that are not strictly legal in character, as long as they contribute to the achievement of the overall objective in question.[6] This approach is well suited to an analysis of the conflict diamonds governance system, which features strong elements of industry self-regulation, and reflects the important role of NGOs and the media in managing the conflict diamonds problem. This reflects the trend towards 'new governance' and 'fragmentation', through which governments are recognised as only one of a number of regulatory agents.[7] The organisation most central to the world's response to conflict diamonds, namely the Kimberley Process Certification Scheme, does not possess formal legal status or impose obligations under the international laws relating to treaties.

Finally, regulatory theory provides insight into criteria that are essential in establishing an effective system in promoting the desired outcomes: the elements of standard-setting, monitoring, and behaviour modification. Through the articulation of particular regulatory approaches and models, regulatory theory provides a further level of sophistication to assist in designing a system that will achieve the desired outcomes.[8]

In terms of regulatory approaches, the two approaches at the opposite ends of the behaviour modification spectrum are command-and-control, and goal-orientated regulation. Command-and-control regulation focuses on punitive action by a central regulator, such as a government agency, which rigorously polices a given industry, and applies punitive measures against those who are non-compliant. By contrast, goal-orientated regulation moves the locus of regulation from the regulator to industry participants, relying on internalised motivations to promote whole-hearted and creative engagement rather than begrudging compliance.[9]

---

6   Parker, C et al. (eds), *Regulating Law* (Oxford University Press, 2004) 1–3.

7   Scott, C, 'Regulating in Global Regimes', Working Paper No 25/2010, University College Dublin (2010) 1–4.

8   Black, J, 'Critical Reflections on Regulation', *Australian Journal of Legal Philosophy*, vol. 27, 2002, 20.

9   Salamon, L M, 'The New Governance and the Tools of Public Action: An Introduction' in L M Salamon (ed.), *The Tools of Government: A Guide to the New Governance* (Oxford University Press, 2002) 15.

More complex regulatory models combine command-and-control and goal-orientated approaches. They articulate particular systems of standard-setting, monitoring and behaviour modification that are designed to maximise the ability of a given regulator to achieve the desired regulatory outcomes. The models presented in this chapter are network models, pyramid models, and hybrid models combining features of both networks and pyramids. These models stand out as being useful in their application to the conflict diamonds governance system, for reasons discussed in greater detail below. Of greatest interest is the hybrid networked pyramid model, as it combines the insights and approaches of both network and pyramid models.

The Network Model

There is an increasing and diverse literature based around the ways in which networks of people, businesses, organisations, and governments act together in a regulatory capacity.[10] Networks contribute expertise and information that assists in carrying out regulatory functions. They may also contribute a range of regulatory interventions to a given system. They typically deploy techniques that are based on dialogue rather than coercion, and are horizontal in the sense that they operate in a non-hierarchical manner.[11] Beyond general principles, more sophisticated models for the regulatory

---

10   For example, see Braithwaite, J and P Drahos, *Global Business Regulation* (Cambridge University Press, 2000) 550–562; Slaughter, A-M, *A New World Order* (Princeton University Press, 2004); Burris, S, P Drahos and C Shearing, 'Nodal Governance' (2005) 30 *Australian Journal of Legal Philosophy* 30; Drahos, P, 'Intellectual Property and Pharmaceutical Markets: A Nodal Governance Approach' (2004) 77 Summer *Temple Law Review* 401; Scott, C, 'Regulation in the Age of Governance: The Rise of the Post-regulatory State' in J Jordana and D Levi-Fuar (eds), *The Politics of Regulation: Institutions and Regulatory Reform for the Age of Governance* (Edward Elgar, 2004); Koechlin, L and R Calland, 'Standard-setting at the Cutting Edge: An Evidence-based Typology for Multi-stakeholder Initiatives' in A Peters et al. (eds), *Non-State Actors as Standard Setters* (Cambridge University Press, 2009); Coen, D and M Thatcher, 'Network Governance and Multi-level Delegation: European Networks of Regulatory Agencies' (2008) 28(1) *Journal of Public Policy* 49; Freiberg, A, *The Tools of Regulation* (The Federation Press, 2010); Mikler, J, 'Sharing Sovereignty for Global Regulation: The Cases of Fuel Economy and Online Gambling' (2008) 2(4) *Regulation and Governance* 383; Williams, C A, 'Civil Society Initiatives and 'Soft Law' in the Oil and Gas Industry' (2004) 36 Winter–Spring *New York University Journal of International Law and Politics* 457.
11   Braithwaite, J and P Drahos, *Global Business Regulation* (Cambridge University Press, 2000) 553–554; Slaughter, A-M, *A New World Order* (Princeton University Press, 2004) 19–20. Issues of power may continue to play a role, albeit less pronounced, in such networks. For example, economic and military strength may be considerations in horizontal diplomacy.

functioning of networks have been developed. Two of these models, discussed below, are the web of dialogue and horizontal government network models.

Close attention to network models and the way they contribute to hybrid networked pyramid models is merited in relation to the conflict diamonds issue, as the Kimberley Process (KP) self-consciously incorporates important features of network governance. Its tripartite structure of government, industry, and NGOs brings together three different networks, each with a particular interest in the resolution of the conflict diamonds problem. Its informal manner of functioning, including the monitoring technique it has labelled as peer review reflect the benefits of networks in creating a regulatory process of socialisation and peer pressure. The manner in which network regulatory models generate insights into the descriptive and normative operation of the conflict diamonds governance system is discussed in depth in Chapter 7 of this book.

Webs of Dialogue

The web of dialogue model was proposed by Braithwaite and Drahos in the context of systems of business regulation at the international level. Given the focus of the Kimberley Process on the regulation of the international trade in rough diamonds, the model has a clear applicability at face value. Apparent in the label, web of dialogue is the idea of a network of different persons and organisations that act on each other using techniques of dialogue to achieve a regulatory purpose. These webs, which often link industry, government, and civil society participants, are identified as the principal means by which regulatory systems are developed and exported across the globe. A web of dialogue may refer to the operations of an intergovernmental organisation, a multinational corporation, or an industry or professional association. The term dialogue refers to a range of non-coercive interactions between actors in a regulatory setting, ranging from discussions in intergovernmental organisations, to mutual auditing between subsidiaries of a multinational corporation, and even naming and shaming of irresponsible corporate practices by NGOs.[12]

In their study, Braithwaite and Drahos concluded that the regulatory technique of dialogue was more prevalent and significant than techniques of coercion and reward.[13] An earlier study of over 100 multilateral treaties to which the US was a party, encompassing both national security and business regulatory matters, found very little resort to sanctions. Indeed, few treaty texts even included a formal enforcement mechanism.[14] Beyond their empirical observation that webs of dialogue are successful in achieving regulatory outcomes, Braithwaite and Drahos attempt to explain why this is the case. They suggest the answer lies in a number of factors, which they describe as complex interdependency, normative commitment, modelling, and habits of compliance.[15]

The initial process of problem definition can be crucial to the success of a web of dialogue. During this process, actors are persuaded to take ownership of a global problem by identifying their own interests in its resolution. An interesting example of this is the manner in which different groups, particularly in the United States, rallied together to address the problem of the depletion of ozone gas in the earth's atmosphere. Naturally enough, environmental groups were involved in raising awareness of the issue, noting the increase in harmful radiation as a result of the loss of significant ozone levels. The conservative US Government was not initially in favour of the international process that led to the Montreal Protocol, but was persuaded to come on side

---

12   Braithwaite, J and P Drahos, *Global Business Regulation* (Cambridge University Press, 2000) 553. There is some debate in the literature about whether naming and shaming is a non-coercive technique deserving the title of 'dialogue' or if it is more coercive than this. Elsewhere, it is suggested that naming and shaming of corporate practices may be more a technique of coercion than dialogue: Ayres, I and J Braithwaite, *Responsive Regulation: Transcending the Deregulation Debate* (Oxford University Press, 1992) 22–24. The context of the naming and shaming may be important to take into account. For example, shame or praise in a largely private context of an organisational meeting may be considered less coercive than an aggressive media campaign targeting a particular corporation.

13   Braithwaite, J and P Drahos, *Global Business Regulation* (Cambridge University Press, 2000) 557.

14   Chayes, A and A H Chayes, *The New Sovereignty: Compliance with International Regulatory Agreements* (Harvard University Press, 1995) cited in Braithwaite, J and P Drahos, *Global Business Regulation* (Cambridge University Press, 2000) 556.

15   Braithwaite, J and P Drahos, *Global Business Regulation* (Cambridge University Press, 2000) 553–554.

as a result of the intervention of a major US manufacturer, Du Pont. Du Pont led research and development towards replacement products for the harmful chlorofluorocarbons (CFCs) responsible for the damage to the ozone layer. As a result, it had a keen economic interest in the international prohibition on CFCs, as it was then in a position to become a global market leader. In this manner, the US Government intervened decisively in favour of the Montreal Protocol, which was finalised in 1989. The level of cohesion created in the ozone-implementation network was described clearly in the words of scientist-diplomat Mostafa Tolba: 'What they would implement, and how, has been based on a circle of friends, an ever-growing circle of friends, that has worked tirelessly under conditions of personal trust.' The subsequent success in replacing CFCs with innocuous hydrofluorocarbons has resulted in the global success story of the ozone problem. Today, levels of atmospheric ozone have returned to acceptable levels.[16]

In the example discussed above, civil society, industry, and government found common cause to establish and implement an international regulatory regime. Braithwaite and Drahos provide an explanation for the cooperative behaviour of disparate groups even in the absence of easily identified self-interest. The explanation, labelled complex interdependency, is that actors seeking to find a solution to a regulatory issue are often engaged with each other in other forums in relation to different issues. As a result, there is an over-arching reason to cooperate, even in the absence of clearly identified self-interest that relates to that specific issue or problem. For example, a national government may be persuaded to cooperate in a global environmental regime through the realisation that it wishes to make headway in upcoming trade negotiations with those same countries involved in the environmental negotiation. This approach can develop into a habit of compliance, whereby parties to a regulatory negotiation will favour compliance as their default position.[17]

---

16   Ibid 264–267. Quote is cited in Canan, P and N Reichman, *Ozone Connections: Expert Networks in Global Environment Governance* (Greenleaf Publishing Limited, 2002) 60–61, which deploys a network model to explain the success of that system (see generally 61–100).
17   Braithwaite, J and P Drahos, *Global Business Regulation* (Cambridge University Press, 2000) 550–562.

Braithwaite and Drahos highlight a number of other factors as supporting the efficacy of dialogue as a regulatory technique. For example, parties may be persuaded as to the normative value of a course of action, and can be convinced by the modelling and compliance of actors they consider as their equals. When a number of national governments agree to a new regulatory standard, a type of peer pressure comes into play, influencing representatives of other national governments to give serious consideration to joining the regulatory regime. This type of peer pressure can be reinforced through reporting obligations, which institutionalise praise and shame for parties to a regime. An example of this is where labour ministries appear before the Freedom of Association Committee or the Committee of Experts of the International Labour Organization (ILO) to account for their efforts implementing ILO agreements. According to Braithwaite and Drahos, a sense of professional pride and honour is created, which motivates representatives to complete with each other in relation to their level of regulatory compliance.[18]

Another important feature of webs of dialogue is the ability to share information. By meeting the informational needs of potential participants in such a network, the costs and benefits of compliance are more easily understood, thereby facilitating decision making. Networks are able to provide technical expertise on aspects of the regulatory system to those parties who may have such a need.[19]

Webs of dialogue and their processes of socialisation and peer pressure share a similar conceptual framework with the idea of isomorphism — a Latin word meaning 'same form' — in the regulatory literature. Under this concept, organisations undergo particular pressures to conform to a common standard (for example, international accounting firms may undergo pressure to conform to unified financial reporting standards). The literature discusses three forms of isomorphism: coercive isomorphism, mimetic isomorphism, and normative isomorphism. Under coercive isomorphism, organisations conform to unified standards as a competitive necessity — for example, to compete for international finance. Mimetic isomorphism operates when an organisation adopts a standard to avoid pitfalls (such as fraud scandals) that appear to have been avoided by standard adopters. Peer

---

18   Ibid 555–556.
19   Ibid 555, 562–563.

pressure by professionals operate under normative isomorphism, in particular with a need to ensure that organisations are able to meet common market needs and expectations in relation to their services.[20]

Horizontal Government Networks

In her work *A New World Order*, Slaughter suggests a network model for international regulation. Her model is proposed both on a descriptive level, as being something that is already occurring in the international environment, and on the normative level, as a constructive approach that entails significant benefits for effective international regulation. Her model is informed by a redefinition of the meaning of sovereignty for a modern nation state. In conceptualising and identifying global networks that influence the decisions of national governments, she posits that the modern concept of national sovereignty is more concerned with a state's ability to influence these global networks, rather than that state's ability to exclude the network from having an influence in its national jurisdiction. Government networks are loosely defined by Slaughter to be a pattern of regular and purposive relations among like government units from different nations working together. Slaughter recognises that civil society and industry groups participate in discussions hosted by government networks, but suggests that privileging of government representatives is appropriate for reasons of democratic legitimacy.[21]

---

20   Werner, J R and J Zimmermann, 'The Evolving Post-National Regulation of Financial Reporting' in H Rothgang and S Schneider (eds), *State Transformations in OECD Countries: Dimensions, Driving Forces, and Trajectories* (Palgrave Macmillan, 2015) 73–74.

21   Slaughter, A-M, *A New World Order* (Princeton University Press, 2004) 11, 14, 56, 261–262; Chayes and Chayes cited in Slaughter, A-M, *A New World Order* (Princeton University Press, 2004) 267. There is a small volume of literature discussing Slaughter's model, although much of it is a discussion of issues of the political legitimacy of the actors involved in government networks, such as Howse, R, 'Book Review: A New World Order, By Anne-Marie Slaughter, Princeton, NJ: Princeton University Press, 2004' (2007) 101 *American Journal of International Law* 231. See also Chung, C, 'International Law and the Extraordinary Interaction Between the People's Republic of China and the Republic of China on Taiwan' (2009) 19 *Indiana International and Comparative Law Review* 233; Anderson, K, 'Squaring the Circle? Reconciling Sovereignty and Global Governance Through Global Government Networks' (2005) 118 *Harvard Law Review* 1255; Antal, E, 'Lessons from NAFTA: The Role of the North American Commission for Environmental Cooperation in Conciliating Trade and Environment' (2006) 14 *Michigan State Journal of International Law* 167; Lang, A and J Scott, 'The Hidden World of WTO Governance' (2009) 20 *European Journal of International* Law 575.

Slaughter suggests the notion of horizontal government networks to explain the cooperation between national agencies from different countries aimed at resolving international concerns. She identifies three broad types of networks within the range of horizontal networks: information networks, enforcement networks, and harmonisation networks. Information networks bring together regulators, judges, or legislators to exchange information and collect best practice. An example is the environmental exchange between the Environmental Protection Agency of the USA and Mexico's equivalent, La Procuraduría Federal de Protección al Ambiente, on monetary penalties in enforcement cases, administrative enforcement procedure and the development of programs for criminal environmental enforcement. [22]

Enforcement networks are motivated by the need of government officials to cooperate with other countries to enforce their own laws. Cooperation involves information exchange and assistance programmes. Perhaps the best known enforcement network is Interpol, which is composed of 179 police services internationally, but has no constituent treaty to constitute itself on a formal basis. Criminal intelligence, including arrest warrants, is shared between member nation police forces through Interpol. Other examples include the European Union's criminal enforcement network, known as Trevi. Enforcement networks may also include the provision of capacity building, or technical assistance. For example, technical assistance from the US Securities and Equities Commission, US Environment Protection Agency, and Justice Department and Treasury Department is a significant contribution to capacity building in the areas of competition, environmental and other forms of regulation globally.[23]

Harmonisation networks generally rely on a treaty or executive agreement, and bring regulators together to ensure that rules in a particular area conform to a common regulatory standard. Harmonisation networks have come under particular criticism as undemocratic, because the technical process of harmonising laws generally bypasses the public, and ignores domestic winners and losers

---

22 Slaughter, A-M, *A New World Order* (Princeton University Press, 2004) 14, 56, 264.
23 Ibid 56.

from the process. Harmonisation is often linked to trade agreements such as the World Trade Organization (WTO) agreements or the North American Free Trade Area agreement, but can also be bilateral.[24]

Slaughter suggests a number of features possessed by horizontal government networks: they are a flexible and fast way to conduct the business of global governance; they are able to coordinate and harmonise national government action; they can initiate and monitor different solutions to global problems; they are decentralised and dispersed, and so cannot exercise centralised coercive authority; and they are government actors, and so are responsible to constituencies that will hold them accountable in the same manner as purely domestic activity, even though they may interact with NGOs of civic and corporate nature.[25]

At the normative level, Slaughter identifies a number of advantages to such arrangements, which benefit in particular weak, poor, and transitional countries, including the exchange of information; the development of collective standards; the provision of training and technical assistance; ongoing monitoring and support; and active engagement in enforcement cooperation. Counterparts in more powerful countries are able to reach beyond their borders to try to address problems impacting within those borders.[26]

According to Slaughter, networks create a system of socialisation that develops and enforces standards of honesty, integrity, competence, and independence in performing regulatory functions. The prestige of membership in a network is often enough to give government officials who want to adhere to high professional standards ammunition against countervailing domestic forces.[27]

---

24   Ibid 19.
25   Ibid 11.
26   Ibid 265–266.
27   Ibid 24.

The Pyramid Model

The regulatory pyramid approach, first articulated by Ayres and Braithwaite in 1992, seeks to combine a number of regulatory approaches into a dynamic synthesis. Its initial formulation concerned regulation at the national level. The pyramid model combines elements of both the deterrence or command-and-control approaches, which argue that law must be tailored towards ill-intentioned people who seek to unscrupulously pursue their interests, as well as elements of the compliance and goal-oriented models, which argue that gentle persuasion is the best approach to securing business compliance. Put another way, it combines rational choice thinking, which argues that business always looks to economic self-interest, along with sociological approaches that give credit to the law-abiding and socially responsible instincts of business. Ayres and Braithwaite argue that it is a false dichotomy to have to choose between punishing and persuading, and that regulators need both approaches in their regulatory armoury. It is a vertical approach in that the model also provides for coercive interventions that may be imposed by a regulator acting from a hierarchically privileged position. The privilege may relate to superior legal authority or simply greater power or influence.[28]

Braithwaite argues that an optimally functioning enforcement pyramid strategy will involve the following elements: a tit-for-tat strategy; access to a hierarchical range of sanctions and interventions, which can be set out in the form of a pyramid diagram; and the availability of a highly coercive sanction at the apex of the pyramid diagram.[29] The model lends itself intuitively to the study of the conflict diamonds governance system as it allows for the type of dialogic interaction and socialisation that occurs within the Kimberley Process, but also goes further to provide for more coercive approaches in appropriate cases. Exclusion from the Kimberley Process, or referral to either the UNSC or the ICC, represent escalations that are available within the system.

---

28   Ayres, I and J Braithwaite, *Responsive Regulation: Transcending the Deregulation Debate* (Oxford University Press, 1992) 20–21.
29   Ibid 40–44.

The Tit-for-Tat Strategy

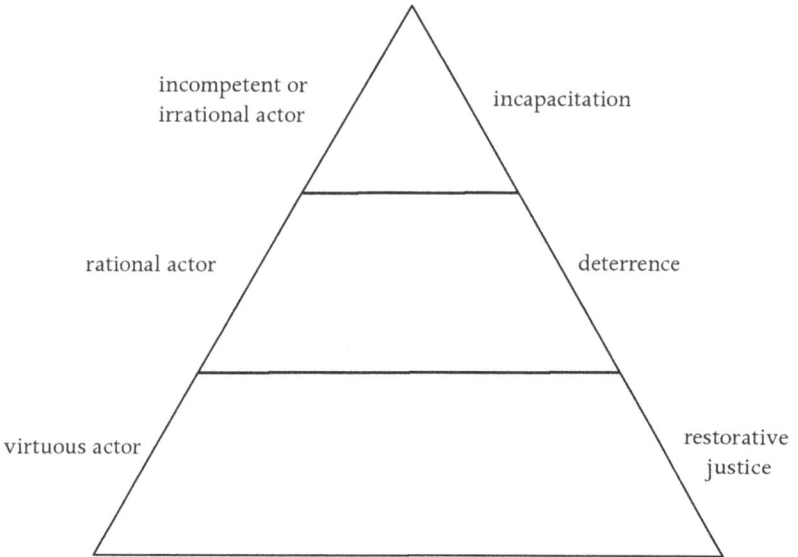

incompetent or
irrational actor

incapacitation

rational actor

deterrence

virtuous actor

restorative
justice

Figure 6.1: Regulatory Pyramid: Escalating Approaches According to Actor

Source: Author's research. Based on Ayres, Ian and John Braithwaite, *Responsive Regulation: Transcending the Deregulation Debate* (Oxford University Press, 1992).

One of the key processes for the functioning of the regulatory pyramid is labelled tit-for-tat enforcement, which can be described as being contingently provocable and forgiving. The fundamental principle of the tit-for-tat approach, which is synonymous with ratchetting up and down the enforcement pyramid, is that it rewards like with like. It starts off with a posture that the regulated entity will be compliant with the regulations. This gives it the advantage of preserving the trust, goodwill, and cooperation of businesses that are seeking to comply with the regulations. A regulator may also employ the technique of positive attribution that encourages long-term compliance — for example, labelling a regulated business as helpful may encourage the directorship of that business to act in a helpful manner. However, in the event that it finds that the business is non-compliant and unresponsive to an approach based on dialogue and persuasion, the regulator gives like in return for like, and escalates one stage up the enforcement pyramid, thereby demonstrating that negative consequences arise from unprincipled business behaviour.[30] The decision by the regulator to start from a position of trust, however, means that the threat of sanctions has low salience for actors who are intrinsically motivated, but can be made salient for those with no intrinsic motivation.[31]

It is important, however, that the minimal sufficiency principle be applied in the deployment of an appropriate sanction to the non-compliant business. A fly should not be hit with a sledgehammer, nor a raging bull with a flyswatter, to use Braithwaite's colourful imagery. By using the minimum sanction necessary to trigger compliance, long-term damage to the relationship between regulator and the regulated business is avoided. When the business returns to a compliant posture of conscientious cooperation, a promptly forgiving posture by the regulator rewards this move and quickly consolidates the desire to return to resource-conserving dialogue and persuasion. It should be noted in this regard that persuasion is cheap, while punishment is expensive, whether in terms of financial or political capital. A strategy based on punishment alone also fosters organised subcultures of resistance, involving regulatory cat-and-mouse, loophole games, and rule proliferation. Punishment damages the capacity of the business to adopt a goal-oriented approach of internalised cooperation.[32]

The core idea of a contingently punitive or forgiving approach is a core idea that is incorporated into the dual networked pyramid model (DNPM) presented in Chapter 7. Under the DNPM, the conflict diamonds governance system is set up in a pyramid fashion to present the idea of a hierarchy of regulatory actors as is required in the regulatory pyramid model of responsive regulation. Under the DNPM, the key regulatory players in the international conflict diamonds governance system are set up in pyramidal fashion, with national governments at the bottom, the Kimberley Process in the middle, the UNSC above that, and the ICC at the apex. The idea of ordering these regulatory actors in this way is to reflect the tit-for-tat, contingently punitive or forgiving concept underlying the regulatory pyramid. For example, the Kimberley Process is able to be contingently punitive or forgiving in relation to a national government that has been found to

---

30   Ibid 21–27.
31   Ibid 49–51.
32   Ibid.

be in contravention of its obligations under the Kimberley Process. While willingness to engage might be rewarded with full membership, and even prestige through positive publicity and recognition, minor or serious non-compliance may attract adverse attention through a critical peer review or even expulsion from the Kimberley Process itself. The ability of the KP to act in this way in relation to a national government, but not the reverse, indicates a hierarchical relationship, which is an understood feature of the regulatory pyramid.

A Hierarchical Range of Sanctions

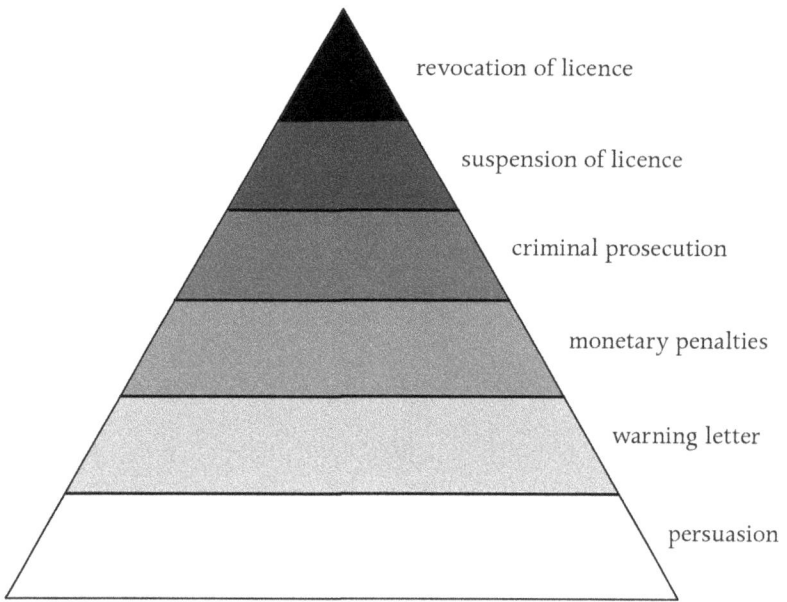

Figure 6.2: Regulatory Pyramid: Escalating Sanctions

Source: Author's research. Based on Ayres, Ian and John Braithwaite, *Responsive Regulation: Transcending the Deregulation Debate* (Oxford University Press, 1992).

Braithwaite argues that businesses will more readily defect from cooperation where a regulator has only one deterrence option, even if that option is cataclysmic. Such a super-punishment is unlikely to be used for political, moral, or legal reasons if it is disproportionate to the seriousness of the crime.[33] For example, prosecutions for criminal

---

33   Ibid.

offences involving long periods of imprisonment are unlikely to be initiated for relatively minor business transgressions, such as creating a low level of pollution.

An example of a regulatory pyramid of sanctions is persuasion — warning letter — monetary penalties — criminal prosecution — temporary suspension of licence — permanent revocation of licence.[34] A discussion of the powers of the Australian Securities and Investments Commission (ASIC) by Assaf provides a real-life example.[35] ASIC has powerful strategies at its disposal, ranging from administrative and civil to criminal options, relating to corporate regulation. Powers possessed by the regulator include freezing assets, and criminal prosecutions can be commenced by ASIC for illegal corporate activity. The ultimate sanction is, arguably, the de-registration of the corporation.

The concept of the regulatory pyramid of sanctions, which sets out key regulators hierarchically, is a core idea in the newly presented DNPM of this book. Each actor in the system has access to an escalating range of sanctions, and there is an ability to ratchet up to a further actor higher up the system who can bring a further range of sanctions to bear. For example, the Kimberley Process is located higher up the DNMP than national governments. The KP has at its disposal an escalating range of sanctions that it can deploy in cases of minor or serious non-compliance. At the lower level, the KP is able to generate critical publicity through a negative peer review. This might be amplified by networked actors such as NGOs who might undertake a negative publicity campaign on the basis of a peer-review report. Further up the scale is the prospect of a national government being expelled from the KP and a blanket export ban on diamonds originating from that country being imposed for serious non-compliance. Further escalations of sanctions may be deployed by other actors in the same system, positioned towards the apex of the DNPM. For example, the UNSC might impose an independent ban on diamond exports acting under its Chapter VII peace and security powers, if particularly concerned by the situation in a country. Beyond this, cases might be referred to the ICC in relation to international crimes committed in conjunction with a conflict diamonds issue, as a further escalation.

---

34 Ibid.
35 Assaf, F, 'What will Trigger ASICs Strategies?' (2002) May *Law Society Journal*, 60, 60–63.

A more general approach to the regulatory pyramid is to describe general tools pitched at the entire industry. An example of a pyramid of broad strategies might be self-regulation — enforced self-regulation — command regulation with discretionary punishment — command regulation with nondiscretionary punishment.[36] A central advantage of this broad strategies approach is that it provides for industry self-regulation as part of the regulatory pyramid, with industry peak bodies effectively acting as an intermediate regulator. In some respects, industry associations can be more important regulatory players than single firms. For example, individual firms will often follow the advice of the industry association to cooperate or face a more interventionist regulatory regime. Peer regulation may also have greater salience to an individual business, as its reputation in the eyes of fellow businesses may be considered more important than its reputation in the eyes of an external regulator.

The escalating systems approach is also reflected in the DNPM, which can be described as a pyramid within a pyramid. The national regulation of conflict diamonds can be considered a distinct pyramid, even while it operates conceptually as a subset of the international conflict diamonds regulatory pyramid. Thus, the idea of escalating regulatory systems is also reflected in the DNPM, albeit that both the national and the international system are stand-alone systems with a full range of sanctions available to them, from the persuasive and self-regulatory to the punitive. It is simply that the peak regulator at the national level, the national government, becomes the regulated entity in the international pyramid, with institutions such as the Kimberley Process and the UNSC taking on regulatory functions.

One of the interesting observations from empirical work in this area is that businesses are intrinsically concerned with adverse publicity, above and beyond simple loss of profits. It has been observed that the personal reputation and corporate reputation of businesses are considered as priceless assets. It follows that a powerful punishment for a business, which can be used in the sanctions armoury of the regulatory pyramid, is adverse publicity.[37]

---

36   Ayres, I and J Braithwaite, *Responsive Regulation: Transcending the Deregulation Debate* (Oxford University Press, 1992) 39.
37   Ibid 22–24.

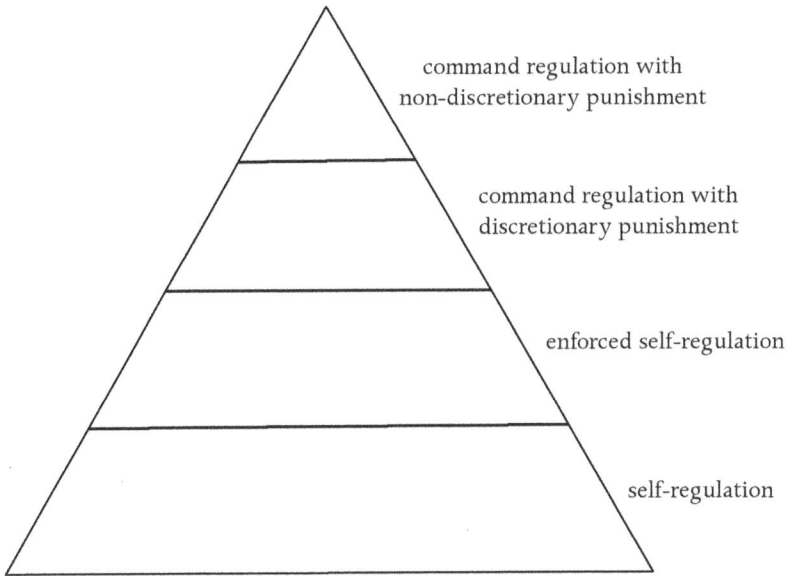

Figure 6.3: Regulatory Pyramid: Escalating Systems

Source: Author's research. Based on Ayres, Ian and John Braithwaite, *Responsive Regulation: Transcending the Deregulation Debate* (Oxford University Press, 1992).

In a more recent discussion of the tit-for-tat strategy, Braithwaite has warned against a rigid, doctrinal use of the pyramid approach. He reinforced the concept that there is a presumption for use of the least coercive technique, but that this is a presumption only. He styles the regulatory interventions as tools in a tool-box, and states that it is important to differentiate these tools from the work of the regulator. That is, the work, or goals, must be in the forefront of the thinking of the regulator. As such, there may be situations where, despite the presumption of starting at the foot of the regulatory pyramid, it is necessary to commence a regulatory intervention part-way up the pyramid, or even at its apex. Braithwaite discusses the possibility that occasionally it is necessary to have a radical escalation or even a radical de-escalation. As an example of a radical de-escalation, he discusses the situation where police confront an armed individual in a siege situation. In such a stand-off, he suggests that a radical de-escalation may involve bringing in the spouse, mother, or other loved one of the armed person, who may be able to persuade that person to surrender.

When discussing interventions by police and other forces in East Timor, he noted that a Catholic nun had intervened to diffuse stand-offs between violent youth gangs.[38]

Availability of Highly Punitive Punishments

Ayres and Braithwaite describe the most effective regulatory players as benign big guns. Such a player, they argue, speaks softly while carrying a very big stick. An example they give is the operation of the Reserve Bank of Australia, which has extensive powers to take over banks, seize gold, and increase reserve deposit rations. However, the Reserve Bank hardly ever uses its powers, but instead relies on persuasion,[39] which becomes a highly effective tool to the extent that it could be dubbed regulation by raised eyebrows. Ayres and Braithwaite argue that the greater the heights of punitiveness to which an agency can escalate, the greater its capacity to push regulation down to the cooperative base of the pyramid. The availability of highly punitive responses helps regulators to cultivate an image of invincibility, so that it is believed that the regulator is good to the loyal, but invincible when it decides to impose sanctions on the disloyal.[40] If the regulator is viewed as being invincible, there is little point in the regulated entity moving for a direct challenge, and cooperation is seen as the only viable approach.

It might also be said that strategic punishment can underwrite regulatory persuasion. In the event of the failure of persuasion, a punitive stance with a recalcitrant company can underwrite the authority of the regulator, who is seen as being fair in the eyes of responsible companies that do not cheat. If the regulatory agency is patient and fair in the escalation, giving warning of the inevitability of the escalation, it enhances even further its reputation for justice as well as strength against recalcitrance.[41] As argued by the Australian Law Reform Commission, effective regulation requires that rules must be implemented in a predictable and consistent manner.[42]

---

38   Braithwaite, J, *Regulatory Capitalism* (Edward Elgar, 2008) 97–104.
39   Ayres, I and J Braithwaite, *Responsive Regulation: Transcending the Deregulation Debate* (Oxford University Press, 1992) 40.
40   Ibid 44–47.
41   Ibid 42–43.
42   Australian Law Reform Commission, 'Penalties: Policy, Principles and Practice in Government Regulation' (Conference Discussion Paper, June 2001).

Tit-for-tat maximises the difference between the punishment payoff and the cooperation payoff. Cooperation is the economically rational response, and where punishment is perceived as a fair response, the intrinsic motivation of the actor continues to be supported.[43] Effective escalation is characterised by a short stick period of discomfort, followed by a longer carrot period of reintegration where the punished party is induced to cooperate with its punishers during the stick period. By inducing cooperation in the stick period, agencies reduce the costs of punishment, and self-punishment moves more quickly onto the carrot phase. In plea bargaining, for example, the threat of stick and stick makes stick and carrot seem the preferable option.[44]

The regulatory pyramid must also have the ability to manage a further category of regulated entities, namely those who act in an irrational rather than self-interested manner. For example, non-compliance may be the result of negligent management of a business rather than wilful pursuit of greater profits. Although mid-level sanctions may assist negligent management to raise its standards (for example, by compulsorily seeking a business consultancy report), the ability to incapacitate a business (for example, by the revocation of its licence to do business) should also be available.[45]

Pyramids with Multiple Regulators in Parallel

Since the original formulation of the regulatory pyramid model in 1992, there has been a significant literature dedicated to testing the theory in numerous empirical settings. This literature has resulted in a number of critiques and re-modelling of the regulatory pyramid approach.[46] One of the significant critiques of the theory is that it underestimates

43   Ayres, I and J Braithwaite, *Responsive Regulation: Transcending the Deregulation Debate* (Oxford University Press, 1992) 49–51.
44   Ibid 43.
45   Ibid 30.
46   For example, Grabosky, P N, 'Discussion Paper: Inside the Pyramid: Conceptual Framework for the Analysis of Regulatory Systems' (1997) 25 *International Journal of the Sociology of Laws* 195; Drahos, P, 'Intellectual Property and Pharmaceutical Markets: A Nodal Governance Approach' (2004) 77 Summer *Temple Law Review* 401; Braithwaite, J, *Regulatory Capitalism* (Edward Elgar, 2008); Braithwaite, J, 'Methods of Power for Development: Weapons of the Weak, Weapons of the Strong' (2005) 26 *Michigan Journal of International Law* 297; Rawlings, G, 'Taxes and Transitional Treaties: Responsive Regulation and the Reassertion of Offshore Sovereignty' (2007) 27(1) *Law and Policy* 51; Parker, C, 'The "Compliance" Trap: The Moral Message in Responsive Regulatory Enforcement' (2006) 40 *Law and Society Review* 591.

the ability of industry self-regulation and third-party regulation to impose coercive measures on recalcitrant businesses. For example, in the diagram showing the pyramid of strategies (Figure 6.3), self-regulation is considered to be at the base of the pyramid, on the basis that it is the most persuasive and least coercive general strategy for business regulation. This was criticised as not being reflective of the potential coercive measures available through self-regulation.[47] For example, the ability of medical practitioner boards to revoke a medical practitioner's ability to practice is the equivalent of the business incapacitation of that individual and is a severe penalty for medical malpractice.[48] Similarly, the NSW legal practitioners' board was recently empowered to suspend the practicing certificates of barristers who went bankrupt as a method for avoiding tax responsibilities.[49]

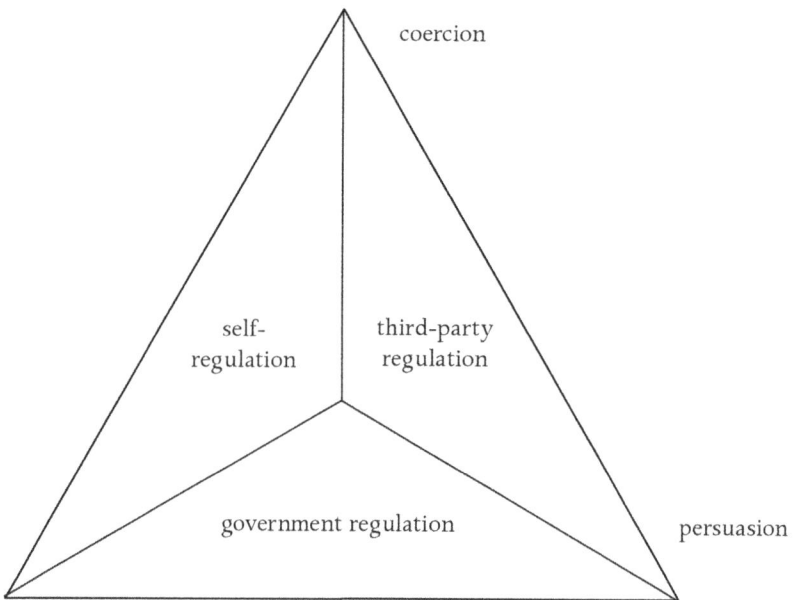

Figure 6.4: Regulatory Pyramid: Multiple Regulators in Parallel

Source: Author's research. Based on Grabosky, P N, 'Discussion Paper: Inside the Pyramid: Conceptual Framework for the Analysis of Regulatory Systems' (1997) 25 *International Journal of the Sociology of Laws* 195.

---

47 Grabosky, P N, 'Discussion Paper: Inside the Pyramid: Conceptual Framework for the Analysis of Regulatory Systems' (1997) 25 *International Journal of the Sociology of Laws* 195, 197–199.
48 Ibid 199.
49 Book, L, 'Refund Anticipation Loans and the Tax Gap' (2009) 20 *Stanford Law and Policy Review* 13–15.

A further critique of the regulatory pyramid was that it did not account for the action of third-party regulators, who are neither governments nor industry self-regulatory mechanisms. Although the original model contemplated action by way of adverse publicity, it did not specifically identify non-governmental or third-party operators as the regulators using this regulatory tool. By contrast, subsequent literature highlighted the ability of such operators to wield powerful coercive tools appropriately located in the upper part of a regulatory pyramid.[50] Examples include the ability of a bank to bankrupt a business in the event of consistent default,[51] and the ability of NGOs to use the media for the purpose of naming and shaming.[52]

In an effort to re-imagine the regulatory pyramid so as to address these limitations, Grabosky proposed a three-dimensional pyramid model (see Figure 6.4) that has three distinct faces. As with the original model, the vertical dimension represents the range of interventions available to a particular regulator, with the most coercive at the apex of the pyramid. Each of the faces of the pyramid represents the efforts of a different regulatory actor: government, industry self-regulation, or third-party regulation. The advantage of having the pyramid appear in three dimensions is that it gives a visual representation of the action of three different regulators, potentially acting simultaneously, in a single model. Each type of regulator is represented as having a full range of possible interventions available in their tool-kit, including highly coercive ones.[53]

Figure 6.4 discusses the concept of multiple regulators in parallel. It demonstrates how different regulators may operate simultaneously on a particular regulated industry, with each deploying a range of measures that range from persuasive to coercive. Operation of multiple regulators in parallel has particular relevance to the DNPM, in particular the tripartite nature of regulation under the Kimberley

---

50   Grabosky, P N, 'Discussion Paper: Inside the Pyramid: Conceptual Framework for the Analysis of Regulatory Systems' (1997) 25 *International Journal of the Sociology of Laws* 195, 199; Scott, above n 497. In Braithwaite, J, *Regulatory Capitalism* (Edward Elgar, 2008) 87–88, he notes that his earlier work discussed 'tripartism', but the role of third-party regulators was not worked seamlessly at that time into the regulatory pyramid model.
51   Grabosky, P N, 'Discussion Paper: Inside the Pyramid: Conceptual Framework for the Analysis of Regulatory Systems' (1997) 25 *International Journal of the Sociology of Laws* 195, 200.
52   Ibid 198–200.
53   Grabosky, P N, 'Discussion Paper: Inside the Pyramid: Conceptual Framework for the Analysis of Regulatory Systems' (1997) 25 *International Journal of the Sociology of Laws* 195, 198–201.

Process, which is depicted as part of the DNPM. Within and beyond the Kimberley Process, industry self-regulation, government regulation, and NGO regulation act simultaneously, with each regulator having at its disposal a range of sanctions starting with those of a more persuasive nature and moving up to progressively more coercive options. For example, within the private confines of the Kimberley Process, NGOs may raise compliance concerns of particular national governments with other parties. They may, however, ratchet up action against a non-compliant government through the means of organising a consumer boycott of diamonds produced by the national government, or indeed organising a boycott of the Kimberley Process. In parallel with regulation by such NGOs, industry and national governments have a range of available persuasive and coercive options at their disposal. Through the instrumentality of the Kimberley Process, industry and national governments cooperate to share information and regulatory approaches and, in cases of serious non-compliance, may recommend, endorse, and enforce a coercive diamond export ban on a particular country.

Pyramids with Multiple Regulators in Sequence

While different regulators may act simultaneously, or in parallel, on a particular regulated group, other systems involve a sequence of independently acting regulators. One such sequence was represented in a regulatory pyramid depicting police action to manage gang-led unrest in East Timor in 2006. At the base of the pyramid were community policing and problem-solving efforts by the Australian Federal Police operating in Dili and other parts of East Timor under international arrangements. Where such efforts were unsuccessful, and police were confronted with organised hostility by gangs, the AFP would pass the baton to the Portuguese elite force called the Guarda Nacional Republicana, who were armed with heavy firearms and had a range of more coercive strategies available to them, including the use of rubber bullets and pushing gangs apart with shields. The apex of the pyramid involved a third group, the joint Australian and New Zealand armed forces, who were able to initiate full-scale military operations as a last resort.[54]

---

54   Braithwaite, J, *Regulatory Capitalism* (Edward Elgar, 2008) 100–104.

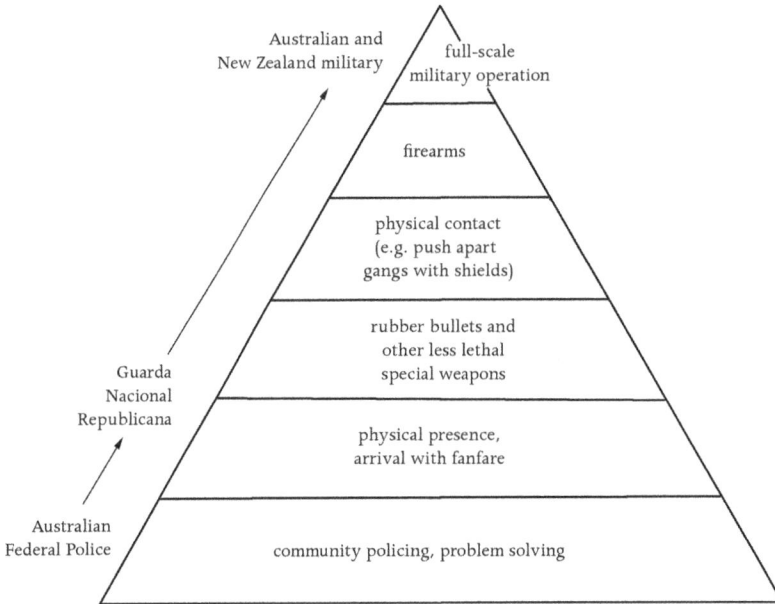

Figure 6.5: Regulatory Pyramid: Multiple Regulators in Sequence
Source: Author's research. Based on Braithwaite, John, *Regulatory Capitalism* (Edward Elgar, 2008).

The concept of multiple regulators in sequence is foundational to the DNPM developed in this book. Such an idea relates to the key idea that regulation is passed from national to international regulators, while, at the same time, the entirety can be understood as a single system for the regulation of conflict diamonds internationally. For example, the central regulator at the national level is the national government, however, national governments are themselves the subject of regulation by international operatives such as the Kimberley Process, the UNSC, and the ICC, particularly in the case that national governments are in serious breach of their responsibilities under the Kimberley Process.

The Strengths-Based Pyramid

A complementary partner to the regulatory pyramid, the strengths-based pyramid, has been proposed in recent regulatory literature. This model, which might also be usefully termed the pyramid of rewards, focuses on rewarding admirable behaviour rather than imposing sanctions on unsatisfactory behaviour. It is the carrot to the regulatory pyramid's stick. Table 6.1 contrasts the two approaches.

While the regulatory pyramid deters through fear and involves risk assessments of regulated parties, the pyramid of rewards creates incentives through hope, and encourages regulators to make opportunities assessments in relation to regulated parties.[55]

Table 6.1: Comparison of Regulatory Pyramid and
Strengths-Based Pyramid

Regulatory Pyramid	Strengths-Based Pyramid
Risk assessment	Opportunities assessment
Fear	Hope
Prompt response before problem escalates	Wait patiently to support strengths that bubble up from below
Pushing standards above a floor	Pulling standards through a ceiling

Source: Author's research. Based on Braithwaite, John, *Regulatory Capitalism* (Edward Elgar, 2008).

Further elaboration of the pyramid of rewards is made easier with reference to Figure 6.6. Each sanctions escalation up the regulatory pyramid is mirrored by an escalation of rewards on the strengths-based pyramid. For example, at the base of both pyramids is education and persuasion, although the focus of such discussion in the regulatory pyramid concerns a problem to be avoided (at pains of possible sanctions), whereas the education in its complementary pyramid is in relation to a strength that is being encouraged. The well-known practice of naming and shaming in the regulatory pyramid is paralleled by naming and faming, through which positive behaviour is praised to encourage the regulated party, as well as bringing the behaviour to the attention of others as a model worthy of emulation.[56]

Ratchetting up the regulatory pyramid are sanctions imposed for failure to meet a standard, so as to deter both the regulated party and others from violating that standard. At the equivalent place in the pyramid of rewards is a prize or grant through which financial reward is added to prestige and praise for achievement and commitment to exceeding minimum standards. Escalated sanctions in the regulatory pyramid lead ultimately to what Braithwaite terms capital punishment, which may also be read, in the business world,

55   Ibid 115–126.
56   Ibid.

as revoking a corporation's right to operate. While it is unlikely that Braithwaite is championing the death penalty for natural persons, his general point is that there should be a serious consequence to either natural persons or business entities that can be deployed in the most extreme cases of non-compliance with regulatory standards. In the parallel world of the pyramid of rewards, the apex might be a highly prestigious and/or financially rewarding prize, such as the academy awards given annually in the motion picture industry. While the apex for either punishment or reward is infrequently bestowed, the possibility of its imposition serves to either deter or inspire, as the case may be, greater action.[57]

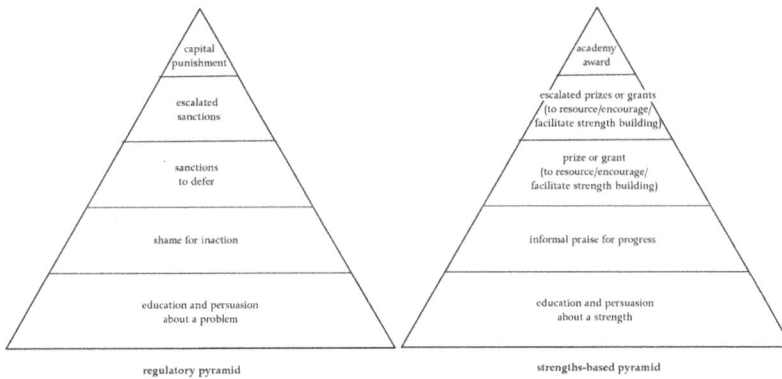

Figure 6.6: Diagram of Regulatory Pyramid and Strengths-Based Pyramid

Source: Author's research. Based on Braithwaite, John, *Regulatory Capitalism* (Edward Elgar, 2008).

The Networked Pyramid Hybrid Model

Over the last few years there have been attempts to combine the essential features of network models and pyramid models so as to benefit from the insights from both theoretical frameworks. In so doing, these new approaches seek to benefit from both the horizontal thinking of the network approaches, as well as the ability to escalate vertically to more coercive forms of intervention. In thinking about

---

57   Ibid.

the simultaneous applicability of two theoretical approaches to a real world issue, it might be recalled that in the field of physics, light, paradoxically, was observed to behave as both a particle and a wave at the same time.

Vertical Networks

Some of the first thinking about mixed models occurred with the development of network models involving elements of coercion. Braithwaite and Drahos proposed a web of reward and coercion. They suggested that reward involves increasing the value of compliance, while coercion is concerned with reducing the value of non-compliance. Techniques of reward include the provision of foreign aid, while coercion can involve economic sanctions or the threat or use of military force. According to Braithwaite and Drahos, only a few actors on the international stage had the resources to deploy reward and coercion techniques, notably the US, EU, China, and the World Bank.[58]

Braithwaite and Drahos argued that webs of reward and coercion were in general less efficient than webs of dialogue and that this was the case because extrinsic pressures overwhelm intrinsic motivation and normative commitment to comply. By contrast, dialogic webs heighten the probability that norms established will be internalised by actors who are part of the web.[59]

Slaughter's vertical government network involved a more sophisticated attempt to bring together network approaches and coercive regulatory interventions, recognising that nation states will, for specific problems, form genuinely powerful supranational institutions that are able to overcome the collective action problems inherent in formulating and implementing global solutions. Hard power is exercised by the institution, which is not simply the combined membership of the network, such as the ability to make a binding decision in relation

---

58 Braithwaite, J and P Drahos, *Global Business Regulation* (Cambridge University Press, 2000) 557–559.
59 Ibid.

to a country, including co-opting domestic government enforcement powers, or excluding a country from membership. This power can be contrasted with the soft power of information, socialisation, persuasion, and discussion. However, as they involve national governments as well as supranational entities, they are still considered to be networks.[60]

The Dispute Resolution Panel of the WTO is the supranational organisation exercising hard power over national governments in the WTO vertical government network. The panel consists of three experts, who make binding decisions based on their understanding of WTO treaty instruments. These decisions affect individual members and the generality of the membership of the government network of WTO members. Other vertical networks are spearheaded by the European Court and the European Commission.[61]

A significant point to be considered in relation to vertical networks is that supranational organisations are more effective in performing functions that states charge them to perform if they can link directly with national government institutions. Such linkages resolve the traditional problem of the inability to enforce decisions of a world body, such as the International Court of Justice, in the absence of a permanent international police force or other enforcement agency. A practical solution to this dilemma is where the existing national enforcement networks are drawn upon by the supranational body. For example, the European Court of Justice interacts directly with national courts to ensure that its decisions are reflected in the decision making of their counterparts at the national level. This is a disaggregated state approach, in which courts interact directly with each other without, for example, being mediated by the respective minister for foreign affairs.[62]

Another example of a vertical network is the complementarity system established by the *Rome Statute for the International Criminal Court 1998*. Under this system, primary jurisdiction is exercised by national courts over war crimes, crimes against humanity, and genocide. It is only where a national court is unable or unwilling to prosecute

---

60 Slaughter, A-M, *A New World Order* (Princeton University Press, 2004) 269.
61 Ibid 13, 269.
62 Ibid 20.

that the ICC may claim jurisdiction and take over a prosecution. The possibility of such a jurisdictional takeover occurring is, in principle, a motivating factor for national prosecutors to take their responsibilities in this matter seriously. It should be noted, however, that the international system benefits from this primacy. The international court is unlikely to have the resources to manage all prosecutions that must be followed throughout the world. Therefore, relying on appropriately well-established national systems significantly relieves this case load, and enables the international court to co-opt the domestic courts to promote its international objectives. National courts, in addition, would increasingly be reliant on precedent-setting cases handed down by the international court, thereby promoting a uniform jurisprudence on international criminal law.[63]

It might be noted that Slaughter does not include secretariats, commissions, and other information agencies under the rubric of vertical government networks, as they are perceived as operating solely though the soft power of information sharing, dialogue, and persuasion only. In this category are placed the technical committee of the International Organisation of Securities Commissioners, the Secretariat of the Convention on the International Trade in Endangered Species, and the Secretariat of the Commonwealth. As they do not possess the binding, coercive powers of bodies such as the European Court of Justice, they are seen rather as handmaidens to national government officials, providing such officials with information needed to coordinate and enforce national law. Nevertheless, this role represents a real level of power, particularly when it is recognised that the professional reputation of member agencies can be buttressed or damaged as a result of compliance information obtained and transferred by a secretariat body. Such modes of operation are increasingly seen as more flexible, responsible and effective than command-and-control approaches.[64]

---

63   Ibid 21, 147.
64   Ibid 156.

Pyramids Linked with Networks and Nodes

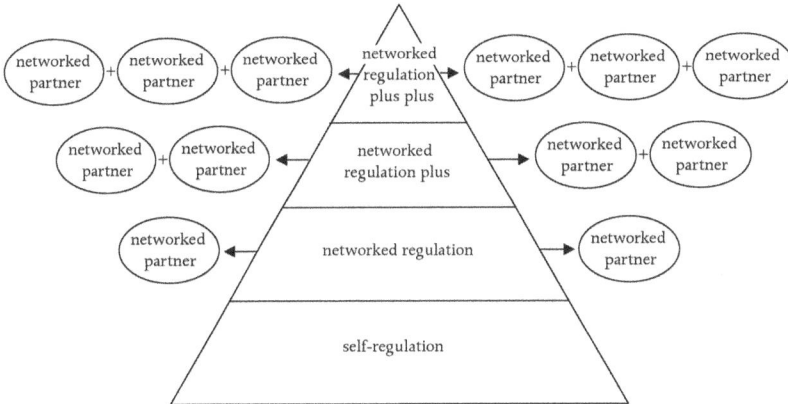

Figure 6.7: Regulatory Pyramid: Escalating Using Networked Partners

Source: Author's research. Based on Braithwaite, John, *Regulatory Capitalism* (Edward Elgar, 2008).

A number of recent developments in regulatory theory have sought to bring together both regulatory pyramid and network models. Some of the concrete examples explored have considered these models in an international context rather than a purely national one. One of the general framework diagrams set out above by Braithwaite shows the ability of a regulator to elicit the support of a new regulator, which is perhaps a network of people, businesses, or organisations. The new regulator adds new resources, information, expertise and regulatory intervention tools that may be deployed. As the resources of further regulators/networks are enlisted, an even greater range of resources, information, expertise and tools are made available to the primary regulator.[65]

---

65   Braithwaite, J, *Regulatory Capitalism* (Edward Elgar, 2008) 94–97.

One of the important developments in networks theory over the last few years is the concept of the node. A node is like the command centre for a network, where resources and expertise are pooled, and key decisions are made. A grass-roots example of a node in relation to a network comes from the movement for peace and security in South African townships. A diverse range of people wanting to promote peace and security through dialogue and discussion at the local level constitutes the network, while the node, where resources are pooled and key decisions made, is the peace committees.[66] The literature also states that there are meta-nodes, where a number of different networks are represented in a single decision-making forum. An example of such a meta-node is discussed below in the section about the international intellectual property regime.

A further model, developed by Drahos, seeks to explain some of the interaction between network and pyramid concepts in a hybrid model. Drahos' model suggests that the reach of a regulatory pyramid is extended by its connection with a greater number of nodes. This model, like that developed by Braithwaite, notes that networks and their nodal command centres add resources, information, expertise, and tools to a primary regulator. The concept of regulatory reach in the context of Drahos' model includes the ability of a regulatory regime, such as intellectual property protection, to operate in an increasing number of national jurisdictions throughout the globe.[67]

---

66   Burris, S, P Drahos and C Shearing, 'Nodal Governance' (2005) 30 *Australian Journal of Legal Philosophy* 30, 30–43; Drahos, P, 'Intellectual Property and Pharmaceutical Markets: A Nodal Governance Approach' (2004) 77 Summer *Temple Law Review* 401, 404–405.
67   Drahos, P, 'Intellectual Property and Pharmaceutical Markets: A Nodal Governance Approach' (2004) 77 Summer *Temple Law Review* 401, 418–419.

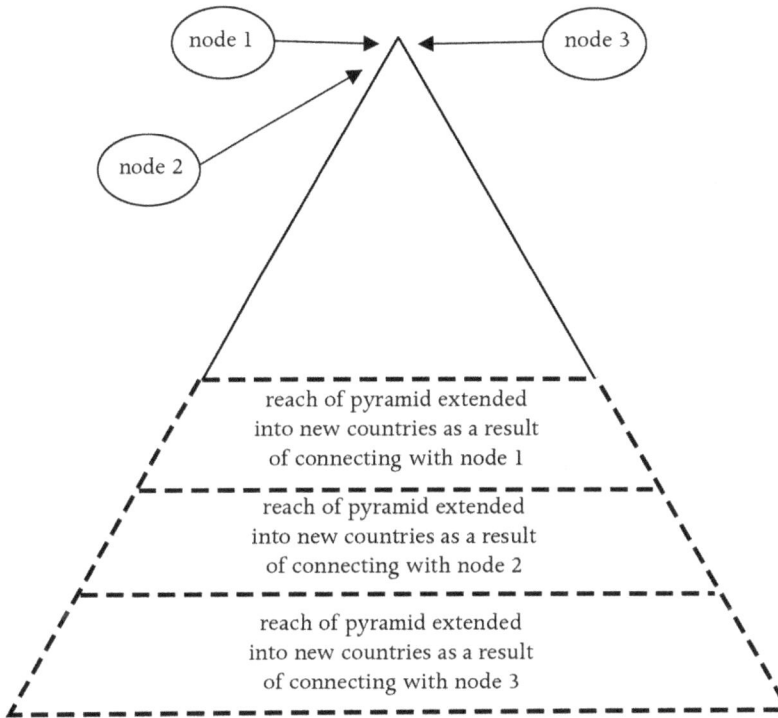

Figure 6.8: Nodes and the Reach of an Enforcement Pyramid

Source: Author's research. Based on Drahos, P, 'Intellectual Property and Pharmaceutical Markets: A Nodal Governance Approach' (2004) 77 Summer *Temple Law Review* 401.

Hybrid Models Applied to Different Systems

The Intellectual Property Regulatory Pyramid

Drahos diagrammatically sets out a significant example of the way in which an international pyramid is constructed, and its utilisation of nodes and networks. The example considers the development and export of legal standards for intellectual property protection from the United States to other countries in the international community. Drahos discusses how the process was initiated and has subsequently been sustained by a large number of multinational corporations based in the United States, which come from industries such as pharmaceuticals, software and entertainment, seeking patent and copyright protection for their products. In particular, the corporations seek intellectual property protection in emerging markets so that they can financially benefit from exporting products or operating there in line with the situation they enjoy in the US domestic market.[68]

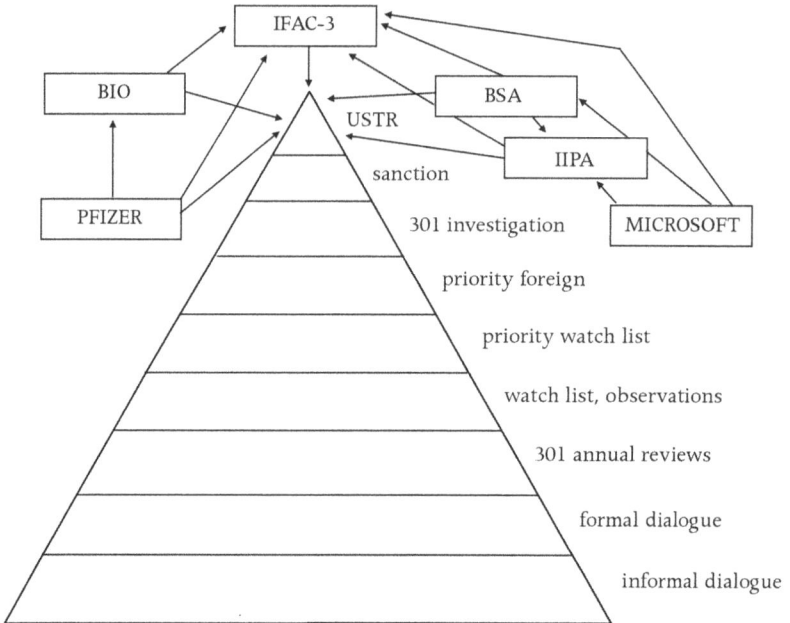

Figure 6.9: Nodally Coordinated International Enforcement Pyramid for Intellectual Property Rights

Source: Author's research. Based on Drahos, P, 'Intellectual Property and Pharmaceutical Markets: A Nodal Governance Approach' (2004) 77 Summer *Temple Law Review* 401.

Drahos charts how the corporations enlisted the support of the US Government in the 1980s, which actively pursued the intellectual property agenda through multilateral treaties such as the Agreement on Trade-Related Aspects of Intellectual Property Rights, and bilateral negotiations. His regulatory model involves both a pyramid, the apex of which is the key regulator, the US Trade Representative, as well as showing connections to the nodes that play a central role in the regulatory system. The interventions available to the US Trade Representative begin with informal dialogue, continue on to listing on various types of watch lists, with the ultimate intervention being the

---

68 Ibid 413–419; see also Burris, S, P Drahos and C Shearing, 'Nodal Governance' (2005) 30 *Australian Journal of Legal Philosophy* 30.

imposition of formal trade sanctions on a country, thereby depriving that country of access to the very large US market for its export goods. The US Trade Representative is empowered by interaction with some important nodes, the most significant of which is the Industry Functional Advisory Committee on Intellectual Property Rights for Trade Policy Matters (IFAC-3). This body is a committee that advises the US Congress and President on all matters that relate to intellectual property in prospective trade agreements. It is composed of a number of representatives of peak industry groups with an interest in international recognition of intellectual property standards, such as the pharmaceutical, software, and entertainment industries. IFAC-3 contributes its expertise, information resources, and political influence towards promoting the goals of US-based multinationals. Besides offering formal advice as to whether an agreement is in the economic interests of the US, the committee occasionally also takes an active role in the finalisation of actual text of intellectual property provisions in agreements. For example, it was a major drafter of the US–Singapore free-trade agreement.[69]

Other nodes identified by Drahos as interacting with the US-based intellectual property regulatory pyramid are peak industry bodies: the Biotechnology Industry Organization, with a membership of more than 1,000 member organisations, the Business Standards Association, which is concerned with copyright issues, and the International Intellectual Property Alliance, with a membership of over 1,100 companies. These are industry peak bodies that are constituted with the goal of advancing the common goals of its membership with respect to intellectual property. Each is directly represented on IFAC-3, although, as Figure 6.9 shows, each may also interact directly with the US Trade Representative. Because IFAC-3 harnesses the resources of a number of nodes, it is described by Drahos as a meta-node.[70]

The concept of the nodally coordinated enforcement pyramid is an important regulatory idea that is developed by the author in the DNPM. In a nodally coordinated pyramid, the regulation is undertaken by a node of actors, in the case of intellectual property regulation,

69   Drahos, P, 'Intellectual Property and Pharmaceutical Markets: A Nodal Governance Approach' (2004) 77 Summer *Temple Law Review* 401, 401–419.
70   Ibid.

business and government. Such nodal action is also reflected in the Kimberley Process, which operates as a node that brings together NGO, corporate, and governmental players. Such players share key information as well as carrying out enforcement action in the case of Kimberley Process non-compliance.

The Traditional Knowledge Pyramid

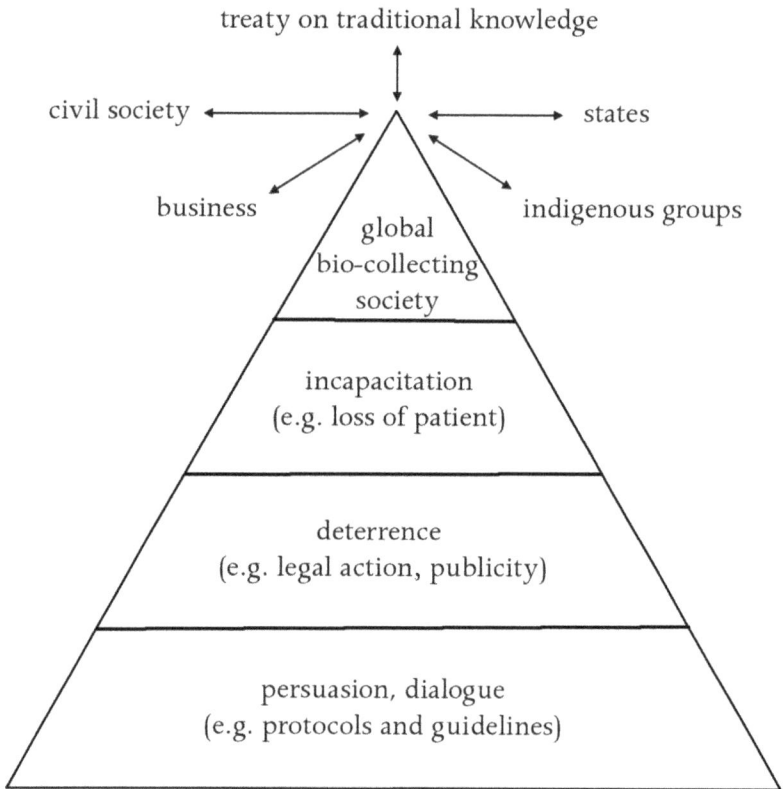

Figure 6.10: International Enforcement Pyramid for Traditional Knowledge

Source: Author's research. Based on Drahos, P, 'Intellectual Property and Pharmaceutical Markets: A Nodal Governance Approach' (2004) 77 *Summer Temple Law Review* 401.

Drahos gives another example of an international enforcement model, involving pyramids and networks, to represent a potential regulatory model for the regulation of traditional knowledge. Traditional knowledge is the broad range of customary knowledge known by indigenous peoples around the world, including the use of genetic resources from local flora and fauna. Drahos proposes a model where the primary regulator is a potential global bio-collecting society. The global bio-collecting society, for example, is well-placed to provide resources to indigenous groups seeking to protect a particular patent. Identifying possible patent infringements by carefully examining patent applications and existing patents can be an exhaustive process, which indigenous groups may not always have the resources to undertake. Therefore, the resources of the global bio-collecting society, and any other groups they may be able to network with such as business, states or civil society, are vital from the regulatory perspective.[71]

The Threat of Collective Debt Default

A further example of an international regulatory pyramid, connected to appropriate networks, has been suggested as a means by which developing countries could leverage their resources internationally around issues of significance to them. The regulatory tool is an example of turning a perceived weakness — the massive scale of the debt of developing countries to countries and financial institutions in the developed world — into a strength. It is based on the somewhat ironic notion that a person with an enormous financial debt to a bank has a lot of effective control over that institution. The concept that developing countries might network for the purpose of making a coordinated default on their bank loans has real weight to it, considering that the collective debt has been estimated to be US$2.5 trillion. Braithwaite suggests that such an intervention might represent the apex of a regulatory pyramid, and that the threat of this intervention could be used by developing countries with a view to making gains in other areas of collective interest. These areas might include the lowering of agricultural tariffs that currently exist in some developed countries against exports from developing nations.[72]

---

71   Ibid 419–424.
72   Braithwaite, J, 'Methods of Power for Development: Weapons of the Weak, Weapons of the Strong' (2005) 26 *Michigan Journal of International Law* 297, 311–330.

Regulation of International Tax Havens

The regulation of international tax havens, also known as offshore financial centres, has also been considered in terms of the regulatory pyramid model. An understanding of the way in which the regulatory pyramid model has been applied to specific international regulatory systems demonstrates the utility of the model in understanding such systems. Such an in-depth understanding is important because the regulatory pyramid model is one of the foundational theoretical models that underpins the author's DNPM.

The central concern of the regulation of international tax havens is the behaviour of a number of nation states, such as Liechtenstein, Monaco, and the Bahamas, which do not impose tax on international persons or businesses, thereby operating as a way for foreign nationals to avoid their tax obligations. The article considered the regulatory efforts of international organisations, including the Organisation of Economic Co-operation and Development (OECD), the International Monetary Fund (IMF) and the European Union. Empowered with significant coercive tools at their disposal, these organisations have attempted to achieve tax law reform in these nations. A further development, however, were attempts of a number of countries to foster bilateral tax arrangements known as double tax treaties. Under the provisions of these treaties, the relevant tax haven undertakes obligations to the effect that, where a person or business had affiliations with another country, it must pay tax either to that country or to the government of the tax haven country. However, according to Rawlings, the effect of negotiating these agreements was to grant recognition and political capital to the tax haven in question. Bolstered by the new political support, it became more difficult for the international organisation to enforce a greater level of cooperation in relation to that country.[73]

The study of international tax havens has utility for the development of a model to help understand the descriptive and normative operation of the international conflict diamonds governance system. The tax havens case study shows that the regulatory pyramid model can be used to hierarchically sequence a number of international regulators as well as the persuasive and coercive interventions they offer. The model

---

73   Rawlings, G, 'Taxes and Transitional Treaties: Responsive Regulation and the Reassertion of Offshore Sovereignty' (2007) 27(1) *Law and Policy* 51, 51–66.

of double tax treaties is a less coercive intervention, as it provides tax relief in the event that tax has already been paid in the country of residence. Potentially more interventionist strategies are applicable through the auspices of the OECD, the IMF, and the European Union. Similarly, the DNPM deploys, in part, the regulatory pyramid approach in an attempt to describe the operation of international and national regulators in the conflict diamonds governance system. Like the international tax havens example, it benefits from the application of the regulatory pyramid, as the pyramid demonstrates how the deployment of an escalating range of sanctions is efficacious in dealing with non-compliance with the conflict diamonds regime. The regulatory pyramid model in the international setting is able to effectively describe the way in which an issue might be transferred to a different international regulator for more efficacious intervention, thereby making use of a sequence of international regulators.

Concluding Remarks

This book considers two main research questions:

1. To what extent has the conflict diamonds governance system achieved its objectives?
2. Does an application of the networked pyramid regulatory model to the system provide descriptive or normative insights into its effectiveness?

The role of this chapter was to provide the necessary theoretical underpinning for a sophisticated response to the second question, which comes from a regulatory point of departure, rather than a strictly legal one. Rather than a purely legal analysis that identifies sources of law, from which particular rights and obligations are derived and articulated, a regulatory perspective considers whether a legal system has been successful in achieving its core objectives. Regulation does not confine itself to strictly legal systems, but includes non-governmental and civil society organisations as protagonists in systems involving standard-setting, monitoring, and behaviour modification.

A number of specific models have been developed to explain why some regulatory systems have proven to be effective, and how their efficacy might be further improved. The network model recognises that regulation occurs horizontally between governments, corporations, and NGOs in webs that promote normative commitment to common goals. The main regulatory techniques involved in such networks are dialogue, persuasion, and socialisation. Although the pyramid model allows for techniques of dialogue, persuasion, and socialisation, it also suggests that there is an important role to be played by more coercive interventions in appropriate cases. If these vertical interventions are employed infrequently and judiciously, then there is an overall reinforcement of the dialogue and socialisation occurring at the base of the pyramid. The networked pyramid hybrid model seeks to combine the key features of both models into a single approach. It recognises, for example, that there may be a network of regulators who operate simultaneously, or who operate in sequence, by passing the regulatory baton to others in the network. Node concept provides a logical connection between network and pyramid models. This concept recognises that networked regulation has a command centre or node that, as well as benefiting from the information-gathering nature of the network, is able to deploy an escalating array of interventions, as envisaged by regulators in the pyramid model.

Recalling that the second research question requires the application of the networked pyramid model to the conflict diamonds governance system, a sub-issue relates to the immediate readiness of the existing theoretical model to accommodate such an application. As discussed in chapters 3 to 5, the conflict diamonds governance system is a complex system involving simultaneous regulatory action at both national and international levels, and in which national governments regulate and are regulated. A system of this complexity has not previously been adequately modelled using the networked pyramid approach. To accommodate these requirements, the model itself needs to be elaborated. It will be argued that an optimal model should include both incentives and sanctions in a single diagrammatic approach. The construction of a model fulfilling these two objectives is addressed in the next chapter.

7

The Dual Networked Pyramid Model: The Pyramid Inside the Pyramid

That which traineth the world is Justice, for it is upheld by two pillars, reward and punishment. These two pillars are the sources of life to the world.

Bahá'u'lláh, nineteenth-century spiritual teacher[1]

Chapter Overview

The second research question being considered by this book is the extent to which the networked pyramid model offers descriptive or normative insights into the conflict diamonds governance system. The previous chapter discussed the state of the art in terms of the networked pyramid regulatory model. The networked pyramid combines the insights of two models, the network model and the regulatory pyramid or pyramid of sanctions model. Network models emphasise the ability of regulators to make significant progress in standard-setting, monitoring, and behaviour modification by networking broadly with diverse groups including non-governmental

---

1   Bahá'u'lláh, *Tablets of Bahá'u'lláh Revealed After the Kitáb-i-Aqdas* (Bahá'í Publishing Trust 1988) 128–129.

organisations (NGOs), corporations and governments and using the soft power of dialogue and persuasion. Beyond this, the regulatory pyramid offers insights into how regulators might respond to unresponsive or antagonistic regulated parties. The pyramid suggests a contingently punitive or forgiving response, in which more coercive interventions are deployed in cases were the soft power of dialogue and persuasion has not provided results.

The insights and approaches of the networked pyramid model are intuitively applicable to the conflict diamonds regulatory system. Nevertheless, two particular issues emerge that arguably suggest a number of modifications to make the networked pyramid a more applicable and powerful model for complex international systems, such as the conflict diamonds governance system. The first issue is that the networked pyramid model, particularly the regulatory pyramid sub-model, was developed primarily in the context of regulation in a national rather than international setting. Particular challenges arise when the networked pyramid is adapted to an international context. For example, are national governments regulating parties, or regulated parties, or both? Can a pyramidal apex be identified as readily in an international context as it can be in a national context?

In fact, these issues arise in any international system involving nation states regulating and being regulated, such as the intellectual property system discussed in the previous chapter. While the intellectual property application effectively demonstrates how nodal coalitions of US-based businesses and US Government instrumentalities deploy a regulatory pyramid approach to foster and enforce the uptake of intellectual property norms by other nation states, it does not simultaneously seek to model the manner in which those intellectual property norms are implemented, monitored, and enforced within those national jurisdictions. Such shortcomings in theoretical modelling are particularly acute in the conflict diamonds governance system because the Kimberley Process explicitly relies on the regulatory role of national governments, while they themselves are regulated by the possible sanction of expulsion from the process, as well as other possible sanctions in the event of UN intervention.

The second issue arises in the context of building the strengths-based pyramid or pyramid of rewards ideas into the networked pyramid model. Typically, this pyramid is modelled separately to the regulatory pyramid or pyramid of sanctions model. A question arises as to whether it is beneficial to model the operation of both rewards and sanctions, or carrots and sticks, in a single diagrammatic model and in so doing whether dynamic interactions or ratchets might become more apparent.

The Pyramid Inside the Pyramid

The networked pyramid model, particularly the regulatory pyramid sub-model, has its origins in modelling regulation in a national context. As such, one of the key regulators is the national government, which will typically apply much of the regulatory apparatus, as well as the sanctions apex of the pyramid: criminal sanctions for individuals or the de-registration of a corporation. The addition of a networked dimension to this analysis, highlighting self-regulation by corporations and naming and shaming by NGOs, can still be conceptualised at this domestic or national level. The immediate challenge of moving to the international level lies in distinguishing the regulating parties from the regulated parties, a question that looms particularly large for national governments. Looking at the bigger picture, it would appear that national governments are regulated parties for the purposes of the conflict diamonds governance system. For example, the Kimberley Process (KP) acts as a regulator, attempting to make sure that national government participants are complying with their obligations, ensuring that only conflict-free diamonds are certified for export, and that only certified diamonds are imported. Regulatory techniques, such as informal naming and shaming through the peer-review system, are deployed with the intention of regulating national governments, as is the highest level intervention of expulsion from the KP for serious non-compliance.

It might also be argued, however, that national governments are just as much regulators in the KP system as they are the subject of regulation. Governments typically rely on the passage of domestic legislation to give them the legal authority to carry out their international obligations under the KP, and are therefore essential in the process of implementing KP requirements. It is the customs apparatus of national governments that must perform the leg-work of actually examining potential rough diamonds exports and issuing KP certificates for those found to be conflict-free. It is national governments that must implement internal controls, such as licensing systems, to ensure that rough diamonds originate from legitimate sources, rather than areas and persons connected with the commission of international human rights crimes. Enforcement options such as fines or criminal prosecutions must be administered at the national level.

In the awareness that national governments act as both regulators and regulated parties in the conflict diamonds governance system, it must be considered whether this double action can be factored into a modified version of the networked pyramid regulatory model. The suggested modification to the networked pyramid model is to embrace the concept of the pyramid inside the pyramid, thereby creating the dual networked pyramid model (DNPM). With reference to Figure 7.1, it can be noted that the bottom layer of the larger pyramid is composed of small pyramids. These small pyramids denote national regulatory systems, in particular national governments. These national governments are seen as being a self-contained networked pyramid unto themselves, as is envisaged in the concept that national governments are regulators in relation to their own jurisdictions. As discussed previously, national governments are charged with implementing their international KP obligations by ensuring that only rough diamonds that have been appropriately certified as being conflict-free are exported.

Figure 7.1 also demonstrates that national governments are the subject of regulation within the context of a larger, international networked pyramid. In this context, the Kimberley Process acts to regulate national governments, applying the regulatory techniques of dialogue, persuasion, informal naming and shaming, and, potentially, expulsion, so as to promote optimal compliance by national governments. The international pyramid also allows for a further range of regulatory interventions in relation to non-compliant national governments. The United Nations Security Council may intervene to impose diamond-trading sanctions on the non-compliant national government. Beyond that, individual members of governments implicated in conflict diamonds practices may be the subject of international criminal prosecution by the International Criminal Court.

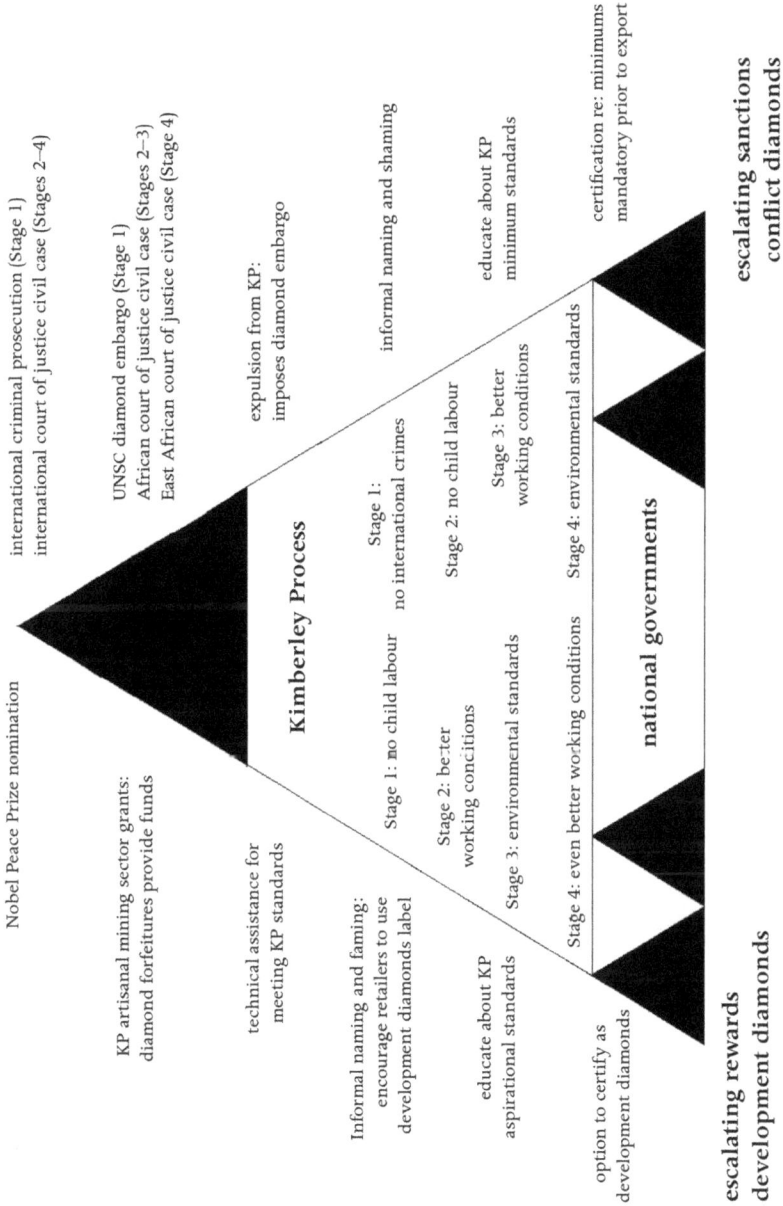

Figure 7.1: The Dual Networked Pyramid Model as Applied to the Conflict Diamonds Governance System

Source: Author's research.

The modified DNPM is therefore a clearly recognisable version of the networked pyramid model as discussed in the earlier theoretical literature. However, it is arguably better able to encapsulate a more sophisticated array of regulatory action than previous efforts to model international systems using the networked pyramid model. In particular, it is able to represent simultaneous regulatory action at the international and national levels using the concept of the pyramid within the pyramid. Such an approach to modelling, beyond application to the conflict diamonds governance system, is of utility to any international system, such as a treaty-based system, where international bodies focus on regulating the implementation of obligations at the international level into the domestic sphere.

Interactive Rewards and Sanctions

A further modification incorporated into the DNPM is the incorporation into the model of the features of the strengths-based pyramid that is also known as the pyramid of rewards. There are, therefore, multiple meanings to the dual nature of the model. It is a duality in the sense that it is a pyramid inside a pyramid, as discussed above, and also in the sense that it combines both rewards and sanctions. The rationale behind incorporating insights from the pyramid of rewards into the model is largely linked to the concept of extending the existing mandate of the KP. While insights from the pyramid of sanctions model may well provide a way forward in relation to the current KP crisis of managing serious non-compliance from Zimbabwe and Angola, it is arguable that the KP, and the system more broadly, might find a new lease of life by extending its mandate to incorporate human rights issues that are of a less serious nature than international crimes. Such issues include the prevalence of child labour in the artisanal mining fields, which deprives the children of the opportunity to undertake a proper education, even where children are working with their parents. As referred to in Chapter 2, child labour is a major problem in cutting and polishing centres in India, with thousands of children missing out on a proper education due to their induction into this industry, which provides only minimal remuneration. Other issues, particularly associated with the artisanal rough diamond mining industry, are poor health and safety conditions, poor remuneration for work performed (less than US$1 a day), and environmental degradation.

NGOs working on conflict diamonds issues have gradually turned their attention to the broader range of problems in the diamond industry. The term 'development diamonds' was coined to denote diamonds that are not merely free from the taint of conflict and international crime, but promote development because they have tackled a broader human rights agenda, such as child labour, health and safety, and remuneration for work performed. The NGO, Diamond Development Initiative International, formed itself around this core concept, and has done important work assessing and comparing the artisanal industries in several countries, and initiating projects to improve conditions on the ground. This book suggests that a further step would be beneficial: incorporating this important work into the core mandate of the Kimberley Process. There would be many significant benefits that would flow as a result of such a move. First of all, the engagement of the organised elements of the KP, namely national governments, corporations, and NGOs, would bring much-needed resources and international attention to this important work. It would also be of benefit to the existing original mandate of the KP, as healthy industrial practices are less likely to degenerate into places that support the commission of international crimes. For example, if an artisanal diamond miner is receiving a good reward for his or her work, that person is less likely to willingly engage with the black market, including conflict diamonds traders, in the hope of better remuneration. On the flip side, and noting that government military and police have been responsible for murder, rape, and beatings in Angola and Zimbabwe, a healthy focus on improving and legitimating the work of artisanal miners would work as something of an antidote to prevailing attitudes that artisanal miners occupy a position very close to the bottom rung on the social spectrum.

A final reason for pursuing the development diamonds agenda through the instrumentality of the Kimberley Process is that the concept of development diamonds lends itself naturally to incentive-based regulation, taking its cue from Braithwaite's strengths-based pyramid. The KP is a certification-based system, which opens the doors to a new type of certification, beyond a simple accreditation that a batch of diamonds is conflict-free. While conflict-free certification is mandatory prior to export, under the terms of the KP certification as a development diamond would not be required prior to export, but would be an option open to the rough diamond exporter. If the rough diamond shipment meets the required criterion to be a development diamond, which might initially be the requirement that it has been produced without the use of child labour, then the exporter would have the option to have it certified as a development diamond at the point of export. The incentive for the exporter to make use of this option is that such certification opens to that person the growing fair trade or ethically produced consumer market in the developed world. As with other fair trade products, there is a market that will pay more for a product with that label, as opposed to a standard product. Ultimately, the intention of the certification is to enable the original artisanal producer to achieve a greater reward from production of a development diamond than a diamond that does not meet development diamond criteria. Thus an incentive is created that encourages the production of diamonds that are in conformity with the relevant human rights standard — for example, the absence of child labour in relation to the production.

One of the innovations of the DNPM is that it brings incentives and sanctions together into a single diagrammatic model, thereby creating a duality independent from the idea of the pyramid inside the pyramid. Although the model is based upon Braithwaite's strengths-based pyramid, the original theoretical discussion of the model took the shape of two pyramids next to each other (see Figure 6.6), thereby contrasting the strengths-based or rewards-based approach with the regulatory pyramid or pyramid of sanctions. Although a compare and contrast approach has its own virtue, combining carrots and sticks into a single model creates a greater capacity to observe the interactions between incentives and sanctions. In particular, there is a facility to clearly model the operation of a regulatory ratchet, whereby aspirational standards may, over time, be recognised as mandatory standards. For example, in the event that the KP embraced a larger mandate, the first stage of operation might involve the elimination of child labour as an aspirational standard linked to development diamonds accreditation, even as freedom from connection to international human rights crimes is required before a shipment may be accredited as being conflict-free and exported. After enough time has passed and considerable progress has been made towards the elimination of child labour, the freedom from child labour standard might be moved from the aspirational to the mandatory side of the ledger. This would mean that export of rough diamonds would not be permissible if they were associated with child labour practices, much as export is not permissible under the current mandate where rough diamonds are associated with international human rights crimes.

There is, moreover, the capacity to ratchet up a further range of human rights standards. These standards would make their first appearance as aspirational goals that would qualify associated rough diamond exports with the label development diamonds. After significant progress has been made towards these goals through the incentive-based framework, and noting the learning process undertaken by industry and regulators in this context, the particular standard might be ratcheted into the mandatory column. Mandatory standards, of course, require compliance as a prerequisite to export. In this manner, a dynamic of progress would be created within the artisanal diamond industry.

As well as creating an interactive dynamic and a ratchetting effect, incorporating insights from the pyramid of rewards into the DNPM allows for easy access to insights from this regulatory model in the context of the conflict diamonds governance system. As provided for in the original strengths-based pyramid model, discussed in Chapter 6, the model provides for increasingly substantial rewards in return for outstanding behaviour that surpasses minimum requirements. The various incentives modelled often mirror available sanctioning responses from the regulatory pyramid. For example, while the regulatory pyramid suggests naming and shaming, the pyramid of rewards provides for naming and faming. This incentive-based process already occurs, to some extent, within the KP as it currently operates. This is because new standards of excellence in implementing KP obligations, including internal controls, are shared with others as a by-product of the peer-review mechanism. The dual model, however, pairs incentives with a self-conscious programme of aspirational standards, thereby allowing for parties to set goals that exceed simple compliance with the original KP mandate. For example, an initiative by an NGO to educate children from the artisanal mining fields, rather than allowing them to be involved in child labour, might be famed at the KP meeting.

A further ratchet up the rewards side of the dual networked pyramid is the possibility of providing grants to NGOs, national governments, or corporations to assist them with projects designed to address the aspirational goals of the suggested new KP mandate. In keeping with the concept of the pyramid within the pyramid is the idea that this segment of the model might itself contain a microcosm of the pyramid of rewards. That is, there might be different levels of grants that could be applied to efforts to improve conditions in the artisanal and cutting/polishing sectors of the diamond industry. One dynamic that might arguably emerge is that an NGO or organisation is provided initially with a grant for a limited duration, or a specific project. In the event that these limited goals are met, or perhaps expectations exceeded, the KP would be likely to provide a larger grant for a longer period of time, creating an upwards dynamic of funding, and progress towards the human rights goals.

Finally, at the peak of the incentives side of the pyramid is a Nobel Peace Prize nomination, chosen for its rough correspondence to international criminal prosecutions, which represents the apex of the sanctions side of the dual model. While any form of international recognition could usefully be employed on the incentives side of the pyramid, a Nobel Peace Prize nomination is arguably the pinnacle of achievement in relation to contributions to world peace, noting that organisations such as the KP arguably would be able to exert influence towards a nomination, with the actual conferral limited, of course, to the Nobel Committee itself. While such a nomination might appear, at first, to be overly ambitious in relation to the fledgling KP, it might be noted that two NGOs working on the issue of conflict diamonds, Partnership Africa Canada and Global Witness, already received such a nomination in 2003.[2]

Concluding Remarks

After considering the question of whether the conflict diamonds governance system has achieved its objectives, this book asks about the extent to which the networked pyramid model might offer descriptive or normative insights into the effectiveness of the conflict diamonds

---

2   Smillie, I, *Blood on the Stone: Greed, Corruption and War in the Global Diamond Trade* (Anthem Press, 2010) 8–9.

system. The networked pyramid model combines the benefits of the dialogic/persuasive networks theory, for responsive parties, with the ability of the regulatory pyramid to escalate punitive interventions in cases of unresponsiveness or outright recalcitrance. The networked pyramid model, intuitively, suggests an ability to describe the conflict diamonds governance system and prescribe improvements to that system. Before applying the model, however, this chapter has suggested two ways in which it might be modified such that it is optimally placed to describe and prescribe in relation to the conflict diamonds governance system.

The first modification relates to the fact that the heritage of the networked pyramid model is largely linked to modelling regulation at the national rather than international level. Recognising that the microcosm of national regulation is semi-independent of the international macrocosm, a modification to the networked pyramid model is suggested, which incorporates the idea of a pyramid within a pyramid. The modified model, which could be labelled the dual networked pyramid model, suggests that the base of the international networked pyramid is comprised of a number of smaller national networked pyramids. In this way, the higher institutions of the international system, in particular the Kimberley Process, act as regulators of the national governments, which are the regulated parties. Under the Kimberley Process, the peer-review mechanism operates by way of informal naming and shaming to influence national governments in the direction of compliance with their KP obligations. However, in implementing their KP obligations within their own national jurisdiction, national governments also act as regulators, involving themselves in standard-setting, monitoring, and behaviour modification in relation to the mining of rough diamonds, and their export.

The second suggested modification creates a further duality in the model, through the combination of insights from the pyramid of rewards, as well as the pyramid of sanctions. Rather than presenting these models as operating in parallel, the DNPM argues that there are synergies to be realised through laying out the operation of both rewards and sanctions in a single model. This interactivity is particularly valuable should the KP decide to broaden its mandate to include a more ambitious human rights agenda. The combination of rewards and sanctions has the potential to create a regulatory ratchet towards higher and more ambitious human rights standards within the industry, particularly in relation to the artisanal mining, and cutting and polishing sectors. The KP as it currently operates sets out minimum standards that must be met before a shipment of rough diamonds can be exported. Sanctions attach to those who export rough diamonds without proper authorisation. By contrast with this approach, this book suggests that the concept of development diamonds can be utilised as a form of voluntary certification that is available to diamond shipments that have been mined and cut/polished in accordance with aspirational human rights standards. While certification as a development diamond is not a prerequisite to export, there is a commercial incentive to do this, as the end product could be labelled as a development diamond at the point of retail, thereby attracting the fair trade or ethical consumer market, who would pay slightly more for a product known to be promoting higher standards of human rights in the industry. It is suggested that a regulatory ratchet might be developed, whereby today's aspirational standard (compliance not required for export) might become tomorrow's minimum standard (compliance required prior to export). For example, the current KP mandate requires that diamonds be free from association with civil war and serious international crime before they can be exported. If freedom from child labour was added as an aspirational goal this year, in several years it might be included as one of the requirements prior to legitimate export under the KP. The ratchet could, furthermore, be used to raise the ceiling in relation to other standards, such as health and safety, remuneration for artisanal diamonds, and environmental standards.

Application of the pyramid of rewards, which escalates in proportion to the achievement above expectations of the parties, might move from informal recognition through naming and faming to grants and awards, with the apex reward a potential Nobel Peace Prize nomination. Financial grants might stimulate progress towards incorporating aspirational standards, and the potential for a high-level prize or recognition serves to raise the height of the aims of parties.

8

Applying the Dual Networked Pyramid Model: Naming, Shaming, and Faming

The Kimberley Process is failing, and it will fail outright if it does not come to grips with its dysfunctional decision-making and its unwillingness to deal quickly and decisively with non compliance.

Ian Smillie, conflict diamonds expert[1]

Chapter Overview

The aim of this chapter is to return to the original research questions. The first of these is an empirical question: to what extent has the conflict diamonds governance system achieved its objectives? In considering this question, it is important to look not only at the degree to which the system has been successful, but the reasons for that level of success, and anything that may be holding the system back from achieving more of its potential. The second research question seeks to provide a theoretical overlay to the conflict diamonds system: to what extent does the networked pyramid model provide descriptive and normative insights into the functioning of the conflict diamonds governance system? It is necessary to provide an appropriate response

---

1   Smillie, I, *Blood on the Stone: Greed, Corruption and War in the Global Diamond Trade* (Anthem Press, 2010) 207.

to the first question so as to be fully equipped to respond to the second question. When the reasons for the relative success or failure of the conflict diamonds system to date are identified, they may be linked to features of the networked pyramid model. In doing so, the networked pyramid theory may provide insight into why the system has been successful, as well as suggesting ways in which it might be improved.

Has the Conflict Diamonds Governance System Achieved its Objectives?

Has the Kimberley Process Achieved its Objectives?

As the centrepiece of the conflict diamonds governance system, the Kimberley Process (KP) has faced a difficult task in seeking to break the link between human rights violations and the rough diamond trade. The central criterion of evaluation is whether the Kimberley Process has prevented the rough diamond trade from benefiting human rights violators. Given the historical lack of transparency that has characterised the diamond industry, it might be commented that there was ample room for improvement. In contrast with this starting position, the Kimberley Process now publishes annual statistical data, setting out the quantity and value of the legitimate trade in rough diamonds, and has come to conclude that conflict diamonds represent less than 1 per cent of the world rough diamond trade. Such important gains are an impressive record for an organisation that is less than a decade old. However, recent challenges have emerged that threaten the ability of the Kimberley Process to achieve its core mandate.

Perhaps the greatest irony of the current challenges to the Kimberley Process is that they do not arise from industry, but rather from government. When the conflict diamonds issue first arose in the 1990s, big business, in particular De Beers, was linked to the illegal trade. However, in subsequent years, De Beers has become a staunch backer of the Kimberley Process. Rather than big business, it has been governmental participants that have thumbed their noses recently at the Kimberley Process. In the cases of Venezuela, Zimbabwe, and Angola, the ongoing integrity and success of the Kimberley Process faces strong challenges. Although human rights violations have not been linked to Venezuelan rough diamonds, its lack of cooperation with Kimberley Process instrumentalities has resulted in a significant quantity of uncertified diamonds entering the world diamond market.[2]

By contrast with its early success with serious non-compliance, the Kimberley Process has been surprisingly spineless in response to these challenges. The Kimberley Process accepted Venezuela's voluntary withdrawal from the process, but has effectively turned a blind eye to the fact that it has recommenced diamond trading outside the Kimberley system. There are several negative implications from Venezuela's action and the Kimberley Process's inaction. One is that many rough diamonds are entering world markets with uncertified origin. Although there has been no concrete evidence that Venezuelan diamonds are linked to human rights abuse, Venezuela's determination not to be involved in the system means that this cannot be clearly verified. It is possible that, while there are no linkages to human rights violations per se, there may be linkages to corruption and bribery, as was found to be the case in neighbouring Brazil. Notably, Brazil cooperated in a review visit by the Kimberley Process and took action to deal with the problems that were unearthed in relation to its diamonds industry. Venezuela also sets a bad example by demonstrating to other nations that it is possible to confront the Kimberley Process and assert your own dominance over it. Finally, the Venezuelan case leaves open the possibility of conflict diamonds entering world markets by being passed off as Venezuelan, and using Venezuela as a conduit place.

A more appropriate response to Venezuela's action would have been to recognise that this behaviour constituted serious non-compliance and for the Kimberley Process to follow-up with expulsion from the process, meaning that other Kimberley members were not permitted to trade with Venezuela while the problem persisted.

The cases of Zimbabwe and Angola are of more serious concern than Venezuela. Turning to the Zimbabwe situation, diamonds deriving from the Marange fields are connected with human rights violations, and are therefore understood to be conflict diamonds.[3] The approach

---

2   The main exception to this analysis is the linkage of De Beers to alleged illegal trading during the early period of the Congolese wars, circa 1998–2002.

3   It should be noted that the definition of a 'conflict diamond' or 'blood diamond' was itself part of the dispute in the Zimbabwe Marange diamonds dispute.

of the Kimberley Process to the emergence of Zimbabwean conflict diamonds has been less than perfect. Rather than ensuring that Zimbabwe was not rewarded for the violence associated with the diamond discoveries, there was a weak response from the Kimberley Process focused on a monitor arrangement near the Marange mine. Beyond this, shipments of rough diamonds have been authorised for sale by the KP, often in controversial circumstances, such as when the civil society coalition walked out from the Kinshasa KP meeting on 23 June 2011. A stronger approach would have been to expel Zimbabwe from the Kimberley Process pending its willingness to ensure that appropriate measures were taken to ensure that conflict diamonds were not able to enter world markets.

In the case of Angola, as detailed in Chapter 2, hundreds of thousands of artisanal miners have been expelled from the Angola to the Democratic Republic of Congo (DRC) in circumstances involving widespread rape, murder, and other human rights violations. Perhaps as a result of the prevailing standoff in the KP regarding Zimbabwean diamonds, the situation in Angola has received very little attention, even from non-governmental organisations (NGOs). However, it would appear that the prevailing situation in Angola might, on its face, represent the commission of crimes against humanity (particularly widespread and systematic rape), meaning that these Angolan alluvial rough diamonds are in fact blood diamonds. Perhaps most ironic of all is the representation of Angola on the KP Artisanal Mining Working Group.

The cases of Zimbabwe, Angola, and Venezuela reveal some important problems in the current manner in which the Kimberley Process manages situations of serious non-compliance. Neither the Zimbabwean situation nor the Venezuelan situation have been clearly identified as cases of serious non-compliance, which is problematic given the threat to the integrity of the system that each represents. There has been no attempt to more clearly define the parameters of what constitutes serious non-compliance. There is a need for further standard-setting in relation to the definition. Unfortunately, the Angolan situation has not been given even this much attention, and remains largely ignored. Perhaps most important of all is the need for a standardised procedure to deal with situations of serious non-compliance. In the event that such situations persist, and the relevant national government is resistant to attempts to bring its behaviour into line with Kimberley standards, it should be accepted that there will be a ratchet up to expulsion from the system. Naturally, there needs to be an ability to reinstate governments that have restored their compliance with the system.

A further problem facing the Kimberley Process is the failure of the consensus method of decision making. As discussed in Chapter 3, this method of decision making is not appropriate for dealing decisively with situations of serious non-compliance. A system of voting requiring some type of majority is required for decisions in such cases. The use of a third-party assessment system would assist the Monitoring Committee in making clear and unambiguous recommendations regarding situations of serious non-compliance.

The Kimberley Process seeks to establish a chain of warranties from the point of production to the point of sale in relation to rough diamonds. Although the primary mechanism established under the Kimberley Process is the export/import certificate, it was always understood that the certificate would be meaningless in the absence of appropriate government controls to ensure the veracity of the certificate. Perhaps most significant in this regard is the ability of producing nations, which are often developing nations with low levels of bureaucratic capacity, to certify that rough diamonds at the point of export are free from association with human rights abuse. An important step in this direction was the decision by the Kimberley Process to make the internal controls of producer countries subject to consideration by review teams. However, further measures need to be enacted. In particular, more detailed guidelines should be developed to identify the main criteria for assessing whether a producer country has adequate internal controls in relation to their rough diamond industry.

Further tightening of the Kimberley Process chain of warranties needs to occur once the rough diamonds have entered countries involved in cutting and polishing, such as India and Israel. This is particularly important given the pervasive use of child labour that characterises the cutting and polishing industry in India, a problem that ought to be addressed by the KP in the context of a new, expanded development diamonds mandate. Ideally, a certificate system should be associated with the production and export of all diamond jewellery, so as to ensure that the originating rough diamonds are conflict-free. Unfortunately, no such protocols have been created. Further down the chain, at the retail end, there is a further lack of implementation. A 2004 study of the practices of jewellery retailers in the US, Canada, and the UK concluded that there was intermittent compliance at best with retail codes of practice. It would appear that customers purchasing diamond jewellery are not given a firm warranty regarding the provenance of the diamonds they are purchasing.

Have the UN Security Council and Tribunals Helped?

As was outlined in chapters 3 and 5, the United Nations Security Council (UNSC) was the first organisation in the international system to gather information about and take action on the issue of conflict diamonds. It actively facilitated the establishment of the Kimberley Process, and continues to provide political support to it through its annual resolutions. Looking further afield, the UNSC was a major player in the establishment of the Special Court for Sierra Leone, which has initiated the first international criminal prosecutions relating to conflict diamonds. The UNSC also has institutional ties to the permanent International Criminal Court (ICC), possessing as it does the ability to make formal referrals to that body. At least one individual is connected to the work of institutions at all three levels. In the late 1990s, Ian Smillie was chosen to serve on the UNSC expert panel, which highlighted the ongoing conflict diamonds problem in Sierra Leone. After this, through his role with Partnership Africa Canada, Ian Smillie was an important influence in the establishment of the Kimberley Process. And it was Ian Smillie who was called on as an expert witness for the prosecution in the Charles Taylor case before the Special Court for Sierra Leone.

Noting the interconnectedness of the main institutions in what I have termed the conflict diamonds governance system, it seems natural to consider them as part of a network that operates to combat the problem of conflict diamonds. The interventions of the UNSC and the international tribunals have arguably contributed to the level of success achieved by the Kimberley Process in breaking the link between the rough diamond trade and the commission of human rights violations.

The first role played by the UNSC is its monitoring role through its expert committees. These committees, which have investigated situations in the African countries that have suffered from conflict and human rights violations as a result of conflict diamonds, have played an important information-gathering role not only for the UNSC but also for other actors in the regulatory community, including the Kimberley Process, the international tribunals, industry, NGOs, and the media. It might appear redundant in the context of the monitoring now undertaken by the Kimberley Process itself, however, as noted in one critical review, the UNSC reports have sometimes picked up on problems in the implementation of conflict diamonds trading bans that were overlooked by the Kimberley Process monitors in reviews of the same country.[4]

The imposition of economic sanctions, including diamond trading sanctions, on countries in breach of their obligations relating to conflict diamonds represents an important coercive ratchet that is available to the conflict diamonds governance system. Such interventions were particularly significant in the period before the establishment of the Kimberley Process, but retain their importance in a multi-regulatory environment. In the event that a member country has been excluded from the Kimberley Process, and therefore excluded from the rough diamond trade through this mechanism, UNSC sanctions represent a more coercive escalation. This is because resolutions imposing sanctions issued with reference to the Chapter VII powers of the UNSC are legally binding under international law. The United Nations resolutions typically address a range of related issues, such as import bans on armaments, and potential bans on diamond exports. The political authority of the UNSC, combined with the possibility of further escalation, suggests that such resolutions have greater impact than exclusions under the Kimberley Process. The UNSC has, in fact, continued its practice of imposing or maintaining diamond trading and other sanctions on African producer countries in particular circumstances, despite the contemplated or actual action by the Kimberley Process. Examples include sanctions on the DRC, Côte d'Ivoire, Central African Republic and Liberia. It is suggested that the ongoing use of this regulatory ratchet has been a continuing factor in the success of the Kimberley Process to date.

---

4   The smuggling of Côte d'Ivoire blood diamonds through Ghana was not picked up by the review visit, but was noted by the UNSC Expert Committee. Smillie, I, 'Paddles for Kimberley: An Agenda for Reform' (Report, Partnership Africa Canada, June 2010) 10.

As discussed previously, the UNSC has important institutional connections with the international criminal tribunal system. The criminal tribunal system has already played an important role in relation to the conflict diamonds in relation to at least two countries: the DRC and Sierra Leone. Following the UNSC expert report about the conflict diamonds problem in the DRC, the prosecutor of the ICC issued public notices highlighting the potential liability of those parties involved in the conflict diamonds trade including, potentially, businesses at the far end of the trade. Prosecutions initiated by the ICC in relation to the DRC have further highlighted the role of conflict diamonds in exacerbating the conflict there, and its connection to grave human rights abuses, including the use of child soldiers and the killing of civilians. More recently, the cases before the Sierra Leone Special Court have provided even greater attention to the problem of conflict diamonds. Conflict diamonds have been used to provide context to finalised cases, prove liability to subsequent human rights violations, and are connected — through mining processes — directly to human rights violations. There was significant media attention to the proceedings of the Special Court, particularly the high-profile prosecution of Charles Taylor, which even involved the testimony of celebrity supermodel Naomi Campbell. The conviction of members of the Revolutionary United Front (RUF) and Armed Forces Revolutionary Council (AFRC) leadership on charges related to conflict diamonds, and the ongoing prosecution of Charles Taylor, have sent a clear message that, in the most serious cases, persons involved in orchestrating the conflict diamonds trade will be subject to criminal prosecution. There is also a moral dimension to prosecution, which serves to reinforce social norms against a trade such as the conflict diamonds trade. The trials before the ICC and the Sierra Leone Special Court have focused international attention on the issue of conflict diamonds and thereby reinforced the important work of the Kimberley Process in seeking to regulate the trade in rough diamonds. Even political leaders of national governments have been made to realise that involvement in the conflict diamonds trade, and commission of human rights abuses, may result in international criminal prosecutions. This has increased general willingness of member nations to cooperate fully with the other international agencies of the conflict diamonds governance system, such as the UNSC and the Kimberley Process.

In turn, such a contribution must be recognised as one of the factors resulting in the reduction of the conflict diamonds trade to less than 1 per cent of the international rough diamonds market.

Applying the Dual Networked Pyramid Model to the Conflict Diamonds Governance System

Network Features of the Conflict Diamonds Governance System

The Network as a Descriptive Tool

The network models, regulatory webs, and government networks provide important insights into the operation of the conflict diamonds legal system, particularly the functioning of the Kimberley Process. The value of networked regulation was seemingly built into the Kimberley Process, through the use of the tripartite government, business, and NGO structure. Each category of participant is, in a sense, a network. The fact that the majority of governments involved in the rough diamond trade are represented shows that Kimberley is, at least in part, a government network. The Kimberley Process was also the catalyst for the large-scale diamond industry to organise itself through the formation of the World Diamond Council (WDC). With the involvement of the WDC at the Kimberley Process, as well as member businesses such as Rio Tinto and De Beers, the Kimberley Process represented a network of rough diamond trading businesses. Given that each NGO is itself a network of like-minded persons, the bringing together of a number of NGOs of global reach through the Kimberley Process is another network feature of the organisation.

The Kimberley Process can also be understood as a node that acts as a command centre by bringing together representatives of the government, NGO, and business networks. The KP processes the information gathered by these expansive networks and makes concrete regulatory decisions in response.

There are a number of specific features mentioned in the network models that correlate closely to the operation of the Kimberley Process. The concept of webs of information is very apt in regard to the Kimberley Process, which brings together expansive networks of information through government, business, and NGO networks. Such information is vital for appropriate regulatory action to occur. For example, in the event there is an outbreak of human rights abuses fuelled by conflict diamonds, it is often the NGO network, which has been coined 'the conscience of the Kimberley Process', which brings it to the attention of the Kimberley Process. The manner in which such information is brought back to the Kimberley Process could also be described in terms of webs of accountability. Importantly, the Kimberley Process incorporates a type of separation of powers, where each power or interest is able to bring accountability to the others. In the event that big business gets caught up in the conflict trade, national governments, which are institutionally removed, may bring them to account, particularly national governments from different countries with no particular stake in that business.[5] Big business has an interest in holding to account parts of the industry, including the artisanal industry, which might bring the diamond trade as a whole into disrepute. Besides highlighting contraventions, big business has an incentive in assisting small business or parts of the trade that have been traditionally troubled in building capacity and becoming Kimberley compliant. Finally, and perhaps most significantly, is the role that NGOs play in promoting accountability. Entering into the Kimberley Process without an economic stake in the rough diamond industry, they are able to present an objective third-party perspective into the operation of the Kimberley Process. In particular, they are in a position to hold to account not only business interests but also national government interests where these diverge from the larger purpose of the Kimberley Process as a whole.

A further insight into the Kimberley Process derives from Ann-Marie Slaughter's government networks theory, which posits that networks of government officials group together around particular subject-matter domains. One of the ways in which they operate she describes as horizontal government networks. Such networks avoid

---

5   Braithwaite, J, 'Realism and Principled Engagement in International Affairs and the Social Sciences of Regulation' (Paper presented at RegNet@10 Conference, The Australian National University, March 2011) 3.

hierarchical structures, but operate on the basis of government officials coming together as equals, giving rise to the operation of peer pressure. Through this type of peer pressure, government officials find themselves in social relationships to other governments, and through this connection feel a sense of accountability for obligations that are mutually undertaken. This model is particularly mirrored in the working of the Kimberley Process peer-review system. In the Kimberley Process peer-review system, government representatives, assisted by NGO and business representatives, review other governments in terms of their degree of compliance with Kimberley Process requirements. One of the main ways in which this system works is that it draws on the sense of mutual obligation by which Kimberley Process governments come together. Through highlighting an inadequacy in the operation of one nation's diamond processing system, pressure is placed upon that nation to come into greater conformity with its obligations.

The Network as a Normative Tool

It is arguable that the success of the Kimberley Process to date may largely be attributed to its ability to incorporate features from the networks regulatory model. However, as discussed below, this book contends that if its operating procedures more closely conformed to features of the pyramid model, further improvement of the conflict diamonds governance system would follow. However, there should be no backtracking of the networks features already incorporated into the Kimberley Process and its related regulators. While the *sine qua non* of the networks theory is socialisation, the central feature of pyramid theory is its ability to deploy coercion in an optimal manner. However, there is a noted risk in the literature around pyramid theory that coercion, particularly as embodied in classic command-and-control theory of regulation, tends to undermine the normative good will and cooperation that are gained through socialisation, dialogue and persuasion. Therefore, it is important in discussing the deployment of coercive interventions to keep in mind that they must be made in a context where the potential to harm the gains made through persuasive functioning are minimised. This challenge is discussed further below.

While the conflict diamonds governance system has significant lessons to learn through analysis in the light of the networked pyramid model, there are arguably insights that might be suggested in relation to networked pyramid regulatory theory that arise out of a study of the conflict diamonds governance system. One such insight relates to networks theory and understanding of the node concept. Nodes are considered the command centres of networks, with meta-nodes involving representatives of several networks. Nodes are the places where information from the various networks is gathered, and where regulatory action is determined and carried out. It appears on its face that the Kimberley Process conforms to this definition. The KP plenary brings together three major networks: representatives of NGOs, industry, and national governments. It is the plenary, and between meetings the chair and committees, which carry out regulatory functions relating to conflict diamonds, including information gathering (annual reports, statistics, review visits), standard-setting through developing terms of reference for working committees, procedures, or suggestions to reform the primary agreement, and behavioural modification activities, such as reporting on non-compliance issues highlighted in review visits, reporting issues to the UN system, or recommending that countries be expelled from the Kimberley Process. The Kimberley Process has the further capacity to engage with other networked regulators in the conflict diamonds governance system. It can receive a ratchet from below from national governments, such as when Côte d'Ivoire requested that it be removed from the Kimberly Process due its inability to manage conflict diamonds. The Kimberley Process is also able to ratchet up regulatory interventions to more interventionist institutions such as the UNSC or even the ICC. In these ways, it functions as a powerful meta-node, much in the manner that IFAC-3 operates as the major regulatory node in the export of US-developed intellectual property rules throughout the international community.

There are, however, significant differences between the nature of the intellectual property node described by Drahos and the nature of the node represented by the Kimberley Process. First of all, the intellectual property node is less formalised than the Kimberley Process. Although there are references to its functioning in terms of finalising trade treaties by the US Congress, the Kimberley Process Agreement is a more comprehensive constitutional document, placing the organisation on a more formal footing than IFAC-3. The other difference, however, is more significant in terms of the nature of nodal theory. In the intellectual property regime, IFAC-3 and the US

Government fundamentally agree on the same agenda, and promote the same interest. This is perhaps made easy because it is concerned with the alteration of rights overseas to the advantage of US corporations, and is therefore not confronting US interests in a direct manner. The nature of this node is very different to the Kimberley Process, which brings together divergent interests in the name of solving a common problem. Prior to the identification of conflict diamonds as a problem, even large corporations seemingly had no problem in purchasing them. National governments do not necessarily identify regulation of this sector as being in their immediate national interest, as demonstrated by the recent approach of Venezuela and Zimbabwe. It follows that the task of forging a common purpose and common agenda between NGOs, industry, and national governments requires looking beyond immediate self-interest in a manner that is not modelled in the intellectual property regime. Indeed, the bringing together of the three disparate groups can be seen as an example of separation of powers. This constitutionalist concept, previously discussed, indicates that, as far as possible, there should be a division of powers amongst different power bases, and not simply in the normal manner of government (i.e. between judiciary, executive, and legislature). Some examples given in the literature include separation of traditional/tribal power from power in terms of the modern state in developing countries, the separation of church and state, and the breaking up of the US military industry complex.[6]

Perhaps the best analogy to the manner in which the Kimberley Process formalises the tripartite involvement of civil society, industry, and national governments relates to the much more venerable International Labour Organization (ILO). The ILO, which seeks to set standards regarding labour standards internationally, combines governments with employers and employees. It is famously one of the only organisations to survive the transition from being connected with the League of Nations to the post-war era of the United Nations (perhaps the only other structure to do so being the Permanent Court of International Justice, which was transformed into the International Court of Justice, although this continuity is debated in the literature). The three separate interests that are brought together in the ILO hold each other accountable as each seeks different outcomes through

---

6   Ibid.

serving the point of view of their differing constituencies. The ILO could certainly be conceptualised as a node according to networks theory, however, it appears that this has not yet been done. Perhaps this study of the Kimberly Process as a node will prompt scholarly investigation into the manner in which the ILO operates in a similar fashion.

Pyramid Features of the Conflict Diamonds Governance System

The Pyramid as a Descriptive Tool

The regulatory pyramid tool is particularly useful in describing the vertical aspect of regulation, that is, the manner in which increasingly coercive measures may be applied to achieve a regulatory outcome. Typically, the regulator who applies such a higher coercive measure is in a more powerful position than the person or entity that is regulated. In considering the application of the pyramid model to the conflict diamonds legal system, it is perhaps most easily seen in terms of pyramids at two different levels: the national level and the international level. At the national level, the primary regulator is the national government and the regulated persons are artisanal and industrial rough diamond miners and traders. At the international level, the Kimberley Process is the initial regulator, with national governments being the regulated entities. Peer review represents a regulatory ratchet available to the Kimberley Process, as does expulsion from the system for serious non-compliance. Further regulatory ratchets in relation to recalcitrant national governments may involve the UNSC and the ICC.

Rather than relying on the earliest iteration of the regulatory pyramid model, this book draws on the latest imagining of the model, particularly as it relates to providing for multiple regulators, both sequentially and in parallel. The model anticipates ratchetting up through the regulatory pyramid, even, potentially, from national to international levels. As such, ratchetting up may pass the regulatory baton from national governments to international actors such as the KP, the UNSC, and the ICC. The incorporation of the idea of regulators working in parallel is a more sophisticated version of the model, which provides a better explanation of the work of multiple regulators on a single regulated entity. The idea that multiple regulators act together with respect to the Kimberley Process incorporates the tripartite nature of the process, which includes, even at the formal level, NGOs, industry, and national governments. Governments, NGOs and industry sectors operate as regulators from within and outside of the Kimberley Process. These three sectors are both regulators and the subjects of regulation. Thus, the subjects of such regulatory activities are industry bodies, other national governments, and civil society bodies.

Descriptively speaking, it is possible to view the conflict diamonds prosecutions as an informal ratchet from the UNSC to tribunal jurisdiction. The two situations involving conflict diamonds prosecutions, Sierra Leone and DRC, have both been presaged by significant action by the UNSC. The UNSC had issued major reports about the role of conflict diamonds in the perpetration of human rights violations in Sierra Leone, and had placed economic sanctions on both Liberia and Sierra Leone. The UNSC was instrumental in the establishment of the Sierra Leone Special Court. The case records, particularly those of the Charles Taylor case, show that the UNSC expert committee reports about conflict diamonds were tendered as evidence, supported by the testimony of one of its experts, Mr Ian Smillie. In contemplating the establishment of the Sierra Leone Special Court, it would be difficult to imagine that its caseload would overlook the role that the conflict diamonds trade played in the conflict.

Similarly, in relation to the DRC cases, the UNSC expert reports highlighted the connection between conflict diamonds and human rights violations in regions such as Ituri, and recommended follow-up action by the UNSC and the international community. It is likely that the UNSC expert reports formed the basis for public statements by the ICC prosecutor regarding the illegality of the conflict diamonds trade. Beyond this, the ICC initiated prosecutions for human rights violations in the area, highlighting the role played by conflict diamonds in exacerbating the situation. In an informal manner, there was a regulatory ratchet between the statements of the UNSC expert committee, its sanctions, and follow-up prosecutorial action by the ICC. Importantly, there is an institutional connection between the two, with the UNSC able to refer cases to the ICC even when such cases would not normally fall within its jurisdiction. Although the institutional mechanism was not formally invoked, it makes sense that the ICC prosecutor would be carefully monitoring expert reports by the UNSC for possible independent action on its own motion.

Descriptively, a focus on the role of national regulatory frameworks within the international system gives rise to the concept of a pyramid within a pyramid. From this point of view, national regulators have a range of interventions available to them in relation to their local diamond industry, with the apex of that pyramid arguably being domestic criminal prosecutions. The prime obligation of the Kimberley Process, by contrast, is to assist national governments in living up to their regulatory tasks. Returning to the national regulatory pyramid, national governments already have a range of regulatory interventions at their disposal. It is clear from analysis of legislation from African producer countries that a number of regulatory tools are already available, including licensing and zoning mechanisms to manage artisanal and large-scale rough diamond mining. They will need to be built upon, as is discussed below, to provide for a better overall approach to the issue of combating conflict diamonds.

The Pyramid as a Normative Tool

One of the key challenges currently facing the Kimberley Process at the international level can be understood clearly in relation to the pyramid model. The challenge of serious non-compliance by Zimbabwe and Venezuela can be seen in terms of a normative application of the regulatory pyramid. The pyramid model requires a hierarchical range of sanctions, with a highly coercive option at its apex. Although the regulatory pyramid was not referred to in its remarks, the suggestions made by Partnership Africa Canada for reform of the Kimberley Process are conceptually very similar:

> In sum, the Kimberley Process needs a rigorous, clear and phased compliance enforcement strategy that starts with assistance and internal pressure, moves to public naming and shaming, and then moves to higher levels of sanctions, suspension and expulsion.[7]

---

7   Smillie, I, 'Paddles for Kimberley: An Agenda for Reform' (Report, Partnership Africa Canada, June 2010) 11.

The Kimberley Process has already been successful at providing technical assistance and internal pressure, especially in the context of its peer-review system. However, it does not have a clear strategy for public naming and shaming, or suspension/expulsion from the Kimberley Process. One of the continued challenges in relation to enforcement via public naming and shaming is the fact that review visit reports are not made publicly available. The simple fact of making such reports publicly available would increase the ability of both the Kimberley Process and civil society organisations to bring the power of adverse publicity to bear on the national government in question. Finally, and most importantly, there needs to be an agreed process for the expulsion of countries in the event of serious non-compliance. As discussed previously, the parameters of what constitutes serious non-compliance need to be more clearly established, with a body exercising a degree of independence able to assess individual cases. While the decisions of this body, operating in conjunction with the Monitoring Committee, might be subject to confirmation by the plenary, as soon as a situation of serious non-compliance is recognised, steps must be immediately initiated leading to a review visit followed by the potential expulsion of that country from the Kimberley Process. In terms of confirmation of the process by the Kimberley Process Plenary, the voting method should provide for some type of majority vote, whether a 50 per cent majority or a two-thirds majority. This would liberate the Kimberley Process from the current deadlock arising from its consensus requirement (which has been interpreted as 'unanimity'), and enable it to take appropriate action.

The pyramid model applied normatively suggests the possibility of further escalation to the UNSC in appropriate cases. There are two possible scenarios where this might be utilised. One is effectively an appeal from the Kimberley Process for a re-categorisation of a case as relating to serious non-compliance. This might be initiated by the Kimberley Process Chair or the Kimberley Process Plenary, following a preliminary decision that a case had been deemed not to have related to serious non-compliance. The UNSC would then have an opportunity to reconsider the situation and, if it was of the view that the original decision was incorrect, it would be able to pass a resolution mandating a trading ban in diamonds with the particular country.

A further scenario is a more traditional use of the pyramid, which would represent a ramping up of sanctions against a particular country that has been expelled from the Kimberley Process for non-compliance. If, in the view of the UNSC, expulsion from the Kimberley Process was not considered as having effectively countered the traffic in conflict diamonds emanating from the expelled country, it could impose, in addition, a trade ban in diamonds mandated under a Chapter VII resolution, or even more extensive economic sanctions (perhaps a ban on arms trading, for example) on that country.

Applying the pyramid in a normative manner suggests that the conflict diamonds governance system would be improved if the regulatory ratchet between the Kimberley Process, the UNSC, and the international tribunals was strengthened. While the connection between the Kimberley Process and the UNSC has been discussed above, there are connections that can be made between the UNSC and the international courts, and even directly between Kimberley and the international courts. One way in which a greater connection could be made between all three levels, even down to the national government level, would be to formulate a specific crime of trafficking in conflict diamonds. The formulation of a formal conflict diamonds trading crime would serve to strengthen the respective regulatory ratchets in several ways. Primarily, it would bring clear subject matter clarity between each regulatory level, which would heighten awareness between regulators of the relationship and strengthen the awareness of potential perpetrators of the regulatory interest at higher levels of the system. For example, although the ICC and the Sierra Leone Special Court have had prosecutions involving conflict diamonds trading, the crimes connected to it have never been labelled specifically as conflict diamonds crimes. Charles Taylor, for example, was charged with rape, the use of child soldiers, and murder and crimes against humanity, with the conflict diamonds trade used as a way of sheeting home criminal liability for the commission of those crimes. While this type of prosecution arguably captures the full extent of his crimes, concurrent prosecution for a more simply defined charge of trafficking in conflict diamonds would arguably have created a stronger conceptual connection to the Kimberley Process and the UNSC. As is the case at present, concurrent prosecution would not result in more serious sentencing in the event of concurrent convictions, as long as the elements of the crime were captured in the alternative conviction.

A conflict diamonds crime before the ICC could be defined in a number of different ways. One definition could attempt to encapsulate the core elements of the definition, namely, trading in rough diamonds in the knowledge that the proceeds would be used to commit human rights violations. However, other definitions could reinforce its connection to other regulators further down the pyramid. In particular, there could be a concurrent definition of conflict diamonds trading as the contravention of UNSC resolutions imposing a ban on trading in diamonds originating from a particular country, along with appropriate *mens rea*. Such an approach would increase the leverage of the key international regulators. For example, Kimberley Process review visits would be able to gather evidence regarding potential conflict diamonds trading in addition to their general functions. The general authority of Kimberley Process review visits would be enhanced by the heightened prospect that an international case may be taken against persons, including government officials and rebel leaders, involved in conflict diamonds trafficking. A Kimberley Process review would be seen much more in terms of the phrase cited by Braithwaite: 'speak softly, but carry a big stick'.

Having an internationally defined conflict diamonds crime would also increase the power of the regulatory ratchet that lies between expulsion from the Kimberly Process and a rough diamond trading ban mandated by a UNSC resolution, as any breach of a UNSC diamonds trading ban would, by definition, constitute a crime. The relevant persons within that country, whether governmental officials or rebel militia leaders, would be on notice that any contravention of that ban would carry a much greater risk of an international prosecution than simply trading despite expulsion from the Kimberley Process. The UNSC would not be required to take formal referral action to the ICC above and beyond imposing diamond trading sanctions on a particular country. The mere fact of its imposing a diamond trading sanction would alert the ICC that its jurisdiction would be activated in the event of a contravention. The ICC could then activate the ratchet by initiating prosecutions in the event of proven contraventions of the UNSC resolution.

In relation to the level of regulation by national governments, the so-called national regulatory pyramid, it is arguable that there need to be further interventions available, both at the base and the apex of the pyramid. At the apex of the pyramid, it is suggested that a clear crime of trafficking in conflict diamonds be enacted. This occurs where the offender knows that the diamonds are connected to the commission of human rights violations. An alternative, perhaps involving a financial rather than imprisonment penalty, is trading in diamonds without due authorisation. This contingency covers the situation where the alleged offender is merely negligent as to ascertain the nature of the diamonds, or even where conflict-free diamonds are traded but in the absence of attending to proper procedures and authorisations.

At the bottom end of the national pyramid are initiatives to assist the diamond industry, the artisanal diamond industry in particular, to be in a better position to be Kimberley compliant. Noting the networked nature of the pyramids being discussed, these initiatives might also connect with the international level. For example, the Kimberley Process established an Artisanal Mining Working Group, involving artisanal miners as representatives. Significant initiatives have already developed momentum, such as the Diamond Development Initiative International, seeking to enhance practices in the diamond industry that promote human rights and better income returns for artisanal miners. These initiatives, spearheaded at the international level by NGOs, need the significant commitment of national governments to be successful. It is perhaps ironic that the governmental actions of Angola, currently heading the artisanal committee at the Kimberley Process, have come under serious criticism on human rights grounds. In particular, the mass deportation of artisanal miners is of serious concern. An artisanal industry that allows for appropriate incomes, in reasonable working conditions, is less likely to go astray to assist rebel movements.

Insights from the Networked Pyramid of Rewards

As has already been discussed, one of the current challenges before the Kimberley Process is a sensible understanding of its existing mandate. It is unfortunate that elements within the KP continue to focus on a narrow reading, which, by focusing on the connection between the rough diamond trade and civil war, would seemingly exclude serious human rights violations from the definition of conflict diamonds. This misunderstanding led the Namibian Chair of the KP, Bernhardt Esau, to make the disturbing pronouncement that 'the Kimberley

Process is not a human rights organisation'.[8] By contrast with this approach, this book argues that the KP is quintessentially a human rights organisation, with the breaking of the link between diamonds and serious human rights violations at the very core of its mandate and the definition of conflict diamonds. In making this point, a further important distinction arises, well understood in the jurisprudence of international law, between serious violations of human rights that qualify as international crimes, such as crimes against humanity, war crimes and genocide, and other human rights breaches that, although they may also be of a disturbing nature, do not qualify as international crimes. The group of crimes that qualify as international crimes are classified in international law as *jus cogens*, meaning that they are fundamental or peremptory norms of international law. This classification has a number of legal consequences. First of all, exceptions to following these rules that may apply to other rules are not allowed, such as because there is a state of emergency at the time: the norm is stated to be non-derogable. Another attendant consequence, in terms of crimes classified as *jus cogens*, is that the obligation to prosecute or extradite attaches. Above and beyond the standard obligation to implement the provisions of a convention, this is an emphatic and specific obligation to either take prosecutorial action through a domestic or international process, or extradite the person to a government or tribunal that will take prosecutorial action. An attendant legal concept, known as *erga omnes,* applies in relation to *jus cogens* and indicates that the international rule is applicable to humanity as a whole. This classification attaches to situations where one country sues another country in a civil international action before a body such as the International Court of Justice. Normally, when international legal action is initiated, procedural rules called rules of standing apply, stating that a case can only be brought by a party that has a legal interest in its outcome. In most cases, this means that the party bringing the action must have suffered in some direct way because of the behaviour of the other party. However, where an obligation is *erga omnes,* any country can take an international law suit, as it is considered that humanity as a whole has suffered a loss as a result of the conduct in question.

---

8   Allen, M, 'The "Blood Diamond" Resurfaces', *Wall Street Journal,* (18 June 2010). Available at: www.wsj.com/articles/SB10001424052748704198004575311282588959188.

It is more accurate to say that conflict diamonds are connected to international crimes than human rights violations for this reason, even though the former is a subset, albeit a more serious subset, of the latter. It is submitted that the connection between international crimes and diamonds is already defined within the parameters of the current KP mandate. The majority of this book has been concerned with improving the effectiveness of the KP to achieve its core mandate, hence the focus on the regulatory pyramid, or pyramid of sanctions, to strengthen its ability to deal effectively with issues of serious non-compliance.

It should be noted, however, that the long-term effectiveness of breaking the link between diamonds and international crimes is linked to the more general condition of the artisanal mining fields. Even in the absence of international crimes, reports on the condition of the informal sector in countries such as the DRC and Sierra Leone show it to be in a deplorable state. As discussed in Chapter 2, problems include the prevalence of child labour, unsafe and unhealthy working conditions, extremely low return for labour (less than US$1 a day, according to one study), and lasting environmental damage to artisanal mining areas. For example, artisanal mining may require extensive periods of digging in mud or dirty water, or spending time in dangerously constructed ad hoc mine shafts.

This book suggests that, in the interests of its own long-term success, the KP vote to extend its mandate to include a broader range of human rights issues and development goals for the artisanal mining sector. To achieve this, the KP might do well to explore the implications of the pyramid of rewards theoretical model. While the regulatory pyramid focuses on sanctions and disincentives, the pyramid of rewards focuses on rewards and incentives. A natural juxtaposition between these two models is to propose, in addition to conflict diamonds, a counterpart that could be termed the development diamond. While the basic concept of the development diamond has already been coined by some of the key NGO players within the Kimberley Process, the main field for work in the area has been removed from the KP itself. An interesting discussion of the politics behind this separation appears in the work by Franziska Bieri entitled *From Blood Diamonds to the Kimberley Process: How NGOs Cleaned up the Global Diamond Industry*. Through her interviews with key players, Bieri shows that while De Beers was prepared to back the upstart NGO Development

Diamonds Initiative International, the remaining voices from industry, represented by the WDC, as well as national governments, were keen to exclude the KP mandate from moving in this direction. A revealing quote from a diamond industry representative put it in these terms:

> This is definitely where the NGOs and the industry are not on the same footing. Some people want to glide the Diamond Development Initiative, which is a fantastically positive initiative that has nothing to do with the Kimberley Process, they want to glide that into the Kimberley Process itself. I think that's a wrong attitude … And in the case of the Kimberley Process we are talking about stopping conflict diamonds from happening, and stopping conflicts in especially diamond producing countries in Africa. We are not talking about free trade, and fair trade. And when you talk about DDI you're talking especially about the fair trade issue. And let's not be forgetting one little tiny detail. We got away with the WTO waiver on the free trade issue, because in fact we are blocking free trade. I am not so sure that WTO would be willing to extend its waiver to cover also issues like fairer trade and DDI … But then of course what you are going to do is you are de facto overloading the scheme, with a fat chance that by doing that the scheme itself will not be very functional any more. And I don't really fancy that.[9]

Here the WDC representative raised a number of concerns about putting the development diamonds initiative into the Kimberley Process. The first concern was in relation to the waiver granted by the World Trade Organization (WTO) permitting trade in conflict diamonds to be restricted, further to exemption provisions under Articles XX and XXI of the General Agreement on Tariffs and Trade (GATT). The GATT establishes a broad framework preventing national governments from imposing restrictions on international trade, although it allows for a number of broadly framed exemptions. Since 2003, the WTO has issued a series of decisions granting countries permission to restrict trade in conflict diamonds. The decisions cite Articles XX and XXI of the GATT, which are broadly framed exemptions concerning measures pursuant to United Nations charter obligations, national security, the protection of human life or health, and public

---

9   Interview with Mark Van Bockstael, 27 September 2005, in Bieri, F, *From Blood Diamonds to the Kimberley Process: How NGOs Cleaned Up the Diamond Industry* (Ashgate Publishing Ltd, 2010) 162.

morals.[10] Considering that the United Nations charter, in articles 55 and 62, refers to the protection of human rights, it is highly likely that the human rights concerns connected to development diamonds and conflict diamonds would be able to come within the auspices of a WTO waiver. Development diamonds issues, such as proper working conditions and freedom from child labour, are human rights concerns protected by international instruments such as ILO regulations and the *Convention on the Rights of the Child*. As well as the more established concerns related to conflict diamonds, the human rights concerns underpinning development diamonds would also justify an exemption under the Articles XX and XXI of the GATT.

Perhaps the stronger argument raised by the WDC representative is that adding development diamonds into the Kimberley Process would overload the organisation and cause it to collapse. There is a logic to this argument, particularly given the challenges that the Kimberley Process has been through regarding Zimbabwe's Marange diamonds. However, some time has passed since the Zimbabwe issue first emerged, and there may now be opportunities not simply to address this question and its related issues (such as the narrow definition of 'conflict diamonds'), but a wider horizon of looking to formally incorporate the development diamonds concept into the KP organisation in order to give new impetus to the organisation. In a recent press release, the incoming KP Chair US Ambassador Milovanovic made a statement about dealing with shortcomings of the KP connected to the evolving nature of conflict diamonds, a reference to unresolved issues arising from the Marange diamonds dispute.[11] It would seem there is a willingness to begin to address this question, even if it is divorced from the actual case of the Marange diamonds. As was discussed previously, one of the challenging after effects of the Marange dispute was disaffection by NGOs, leading at least one high-profile NGO, Global Witness, to formally leave the KP. It may be that a different focus, on bringing development diamonds into the

---

10   The WTO exemption for the Kimberley Process is discussed in Woody, K E, 'Diamonds on the Soul of Her Shoes: The Kimberley Process and the Morality Exception to WTO Restrictions' (2007) 22 *Connecticut Journal of International Law* 335, 349.
11   Kimberley Process, *A Note from Ambassador Milovanovic* (30 July 2013). Available at: www.kimberleyprocess.com/en/note-ambassador-milovanovic.

fold, could reinvigorate NGO interest in the KP, and be a cause for reinvigoration of the organisation, helping it to tackle the more thorny issue of the definition of conflict diamonds.

It might be noted that, even if development diamonds are not formally incorporated into the Kimberley Process organisation, there already exists a level of synergy between development diamonds initiatives and the KP. As envisaged in my dual networked pyramid model (DNPM), upward ratchets modelled on the pyramid of rewards concept might be developed more fully in the current global system. Even in the absence of formal incorporation into the KP, this might be achieved, with reference to my model, through graduated certification that diamonds have achieved particular human rights standards in the domain of development diamonds: free from the use of child labour, in compliance with reasonable working conditions, etc. Such formal certification might be standardised and harmonised, perhaps using particular international benchmarks (such as the International Standards Organisation system). As envisaged in the DNPM, a positive upward ratchet can be created when there is a systematic process of certification from one level of development diamonds compliance to the next. If the process were centralised in the KP it would be stronger and more powerful, as the DNPM seeks to enhance the upward ratchet of the pyramid of rewards with a rising minimum standard. For example, at some point in the future, the KP could declare that it is not permissible to trade in diamonds that were known to have been mined or polished using child labour. Such a trade ban would require the intervention of the Kimberley Process (or an organisation with similar powers) rather than relying solely on aspirational voluntary standard certification.

Measures put into place in Canada reflect aspirational standards in relation to conflict-free diamonds and development diamonds. In publishing a press release on these measures, the Kimberley Process acted to fame Canada for implementing measures that are above compliance with minimum KP standards. This is an example of the pyramid of rewards in effect, as it creates upward momentum for other countries to follow suit. Through calling attention to the achievement by Canada, the KP increased the prestige of Canada, which creates a level of peer pressure for others to follow. The particular measures put into place by Canada introduced a unique standard of excellence in developing a particular system of tracking individual diamonds to particular diamond mines. The centrepiece of this tracking system is a unique inscription on each diamond that is invisible to the naked eye but identifies the mine of origin of the diamond. Canada has also introduced development diamond initiatives that guarantee locals in the area of the mine benefit and the mine conforms to high environmental standards.[12]

Role of WTO in the Conflict Diamonds Governance System

As long ago as 2006, a government representative from Australia was speaking about Kimberley Process winding up due to the completion of its mandate. These words, it would seem, were a little premature, given the emergence of new sources of conflict diamonds from Côte d'Ivoire, Zimbabwe, and Angola at around this time, as well as the ongoing problems in the DRC. What it speaks to, it would seem, is a lack of political will on the part of governments to ensure an ongoing role for the KP. Even if conflict diamonds were no longer entering the global system, there ought to be a process for preventing this occurring, should the nexus between international crimes and the rough diamond trade recur: a preventative role is important. The potential uptake by the KP of a broader human rights agenda is a reinforcement of this preventative role, through addressing the root causes of the emergence of conflict diamonds. Put simply, if more artisanal miners are able to make a reasonable living, in good conditions, without sending their children to work, then they are less likely to sell the product of their labour to benefit international war criminals.

One of the ironies relating to the pyramid of rewards concept is the unfortunate fact that incentives-based approaches have been misused by the international community to exacerbate conflict in central Africa. In particular, the Ugandan and Rwandan governments received inappropriate recognition following their alleged plunder of the natural resources of the Congo, including Congolese diamonds. It was reported that the International Monetary Fund and the World

---

12 Kimberley Process, *Diamond Store Guarantees Ethical, Conflict-free Canadian Diamonds* (Press Release, undated). Available at: www.kimberleyprocess.com/en/system/files/documents/Diamond%20Store%20guarantees%20ethical,%20conflict-free%20diamonds.pdf.

Bank actually praised the Rwandan and Ugandan governments for their unexpected increase in GDP that was, according to UNSC reports, based on plunder of the DRC's natural resources.[13]

Concluding Remarks

This chapter has returned to consider the original research questions:

1. To what extent has the conflict diamonds governance system achieved its objectives?
2. To what extent does the networked pyramid model provide descriptive and normative insights into the functioning of the conflict diamonds governance system?

In responding to the first question, the first sub-question was whether the KP had achieved its objectives. Noting the distinctive collaboration between governments, civil society, and industry that constitutes the Kimberley Process, it has managed in the first instance to socialise the large-scale industry players into being KP supporters. Beyond this, the quantity of conflict diamonds in the international system is at low levels, compared with the extremes of the diamond wars of the 1990s. It is, however, in the area of managing serious non-compliance that the KP now faces its most trying test. Although it was precocious in dealing with the threat posed by diamonds being funnelled through the Republic of Congo-Brazzaville back in 2004, subsequent challenges of equal or greater severity from Venezuela, Zimbabwe, and Angola have been mishandled. In particular, the way in which the KP has turned a blind eye to gross human rights abuses in the Zimbabwe and Angolan artisanal fields shows that it is failing in its core mandate, and raises the risk of the complete collapse of the current arrangements.

The second sub-question considers the extent to which the UNSC and the international tribunal system have contributed to the effectiveness of the conflict diamonds governance system. The UNSC has played a decisive role on the issue of conflict diamonds and was in large part the

---

13   Montague, D and F Berrigan, 'The Business of War in the Democratic Republic of Congo', *Dollars and Sense Magazine* (July/August 2001). Available at: www.thirdworldtraveler.com/Africa/Business_War_Congo.html.

midwife of the Kimberley Process, calling for its implementation in its resolutions. The UNSC has also embodied important monitoring and enforcement roles through its expert committees to various affected diamond-producing countries, and the imposition of diamond trading and other related embargoes. Finally, the emerging jurisprudence on conflict diamonds from the Sierra Leone Special Court and the permanent International Criminal Court has provided support to the work of the UNSC, the KP, and national governments. By prosecuting key leaders in the AFRC, RUF, Civilian Defence Forces and Taylor cases, the Sierra Leone Special Court has sent a clear message that behaviours such as using child soldiers, and terrorising, murdering, and raping civilians in the pursuit of diamond profits are intrinsically criminal and unacceptable to the international community. This message is not lost on the participants in the Kimberley Process, who are galvanised by the prospect of breaking the link between diamond profits and international crime through the chain of custody system.

The second research question considers the conflict diamonds governance system in the light of the DNPM. The analysis is facilitated by breaking the model into its three components, so as to consider insights based on the network model, the regulatory pyramid model, and, finally, the pyramid of rewards model. The essence of the networks model is its reliance on the techniques of dialogue and socialisation to achieve its purposes. The Kimberley Process can be considered as a command centre, or node, where networks from civil society, national government, and the diamond industry engage in dialogue and socialisation. Its successful features largely reflect the benefits described in this model, particularly as seen in the strong engagement and commitment of diamond industry major players to the KP. Its use of peer review is another horizontal technique that has been used to promote best practice in different countries involved with the KP. It is with reference to the regulatory pyramid, rather than networks theory, that the KP's shortcomings are highlighted.

The regulatory pyramid provides for a range of more coercive interventions where particular regulated parties actively oppose the purpose of the regulatory regime. In terms of the functioning of the KP, expulsion from membership represents a more coercive escalation for situations where a government member is in serious non compliance with KP standards. It is argued that the KP should have taken this step in relation to Venezuela, Zimbabwe, and Angola, all of which have actively resisted the core purposes of the KP. This action is important to protect the credibility of the KP, as the failure to remove diamonds originating from these countries from general circulation shows that the conflict-free label provided by KP certification cannot be entirely trusted. The development of a new international crime for trading in conflict diamonds would further reinforce the pyramid structure of the conflict diamonds governance system, particularly as the jurisdiction of the International Criminal Court would automatically be triggered following a breach of UNSC diamond trading sanctions. The enactment of this crime would further strengthen the negotiating hand of the KP, as the consequences for serial conflict diamonds offenders would loom larger.

The pyramid of rewards offers a further window of opportunity for the development of the conflict diamonds governance system. Beyond dealing with the current midlife crisis it finds itself in, the KP and its collaborators have an opportunity to broaden their horizons in the direction of an incentive-based system focused on the concept of development diamonds. Building on the important work of the Diamond Development Initiative International, the KP has a great opportunity to engage with the human rights issues associated with alluvial diamond mining in a proactive, incentive-based manner, beyond its existing sphere of activity. The book proposes that a system of voluntary certification, aimed at connecting to a fair trade niche market in developed countries, accompany the existing system of KP minimum standards. The voluntary certification is made with reference to a number of proposed aspirational standards, which move beyond the core domain of the existing KP mandate to encompass other issues confronting the artisanal industry: child labour, workplace health and safety standards, appropriate remuneration for work, and environmental standards. When both systems are established, the intention is that a standards-raising regulatory ratchet be created, with today's aspirational standard becoming tomorrow's mandatory standard, without which export is denied. An application of the pyramid of rewards to the conflict diamonds governance system might well reinforce the new aspirational standards with a system of naming and faming to parallel the naming and shaming in the regulatory pyramid. That is, countries, corporations and NGOs making particular progress towards aspirational standards might be singled out for particular encouragement. Beyond this, a system of grants might well be established so as to fund new initiatives towards meeting aspirational standards. The apex of the pyramid of rewards — a Nobel Peace Prize nomination — would serve as a fitting counterpart to the apex of the regulatory pyramid — international criminal prosecution.

9

Did You Hear Something?: Concluding Remarks

High on the Jungfrau overlooking the town, it had snowed during the night, and for a few moments the clouds broke to reveal the mountain, looming over the town in its brilliant cloak of new white snow. Unresolved issues notwithstanding, it seemed like a metaphor for the event: a brief opening and a small step towards solving a problem that a group of NGOs had been battling for more than four years.

Ian Smillie, on the establishment of the KP in Interlaken, 2002[1]

Chapter Overview

This chapter summarises the main findings of the book in response to the original two research questions, and sets out recommendations for improving the conflict diamonds governance system based on these findings, directed towards national governments, non-govermental organisations (NGOs), the diamond industry, the Kimberley Process (KP), the United Nations Security Council (UNSC) and the International Criminal Court (ICC). The chapter then suggests areas for further research before giving concluding remarks.

---

1   Smillie, I, *Blood on the Stone: Greed, Corruption and War in the Global Diamond Trade* (Anthem Press, 2010) 191.

Findings

In the first chapter of this book I posited two research questions:

1. Has the conflict diamonds governance system achieved its objectives?
2. Does the networked pyramid regulatory model provide descriptive and normative insights into that system?

Chapter 2 of the book outlined the conflict diamonds problem, showing the connection between the rough diamond trade, conflict, and serious human rights violations in five African producer nations. Chapter 3 introduced the Kimberley Process as the centrepiece of the conflict diamonds governance system, with its import/export certification process established and monitored by industry, government, and civil society representatives. Chapter 4 discussed the operation of the Kimberley Process at the national level. Internationally, the UNSC and the criminal tribunals, the conflict diamonds governance system institutions discussed in Chapter 5, have also played key roles through imposing economic sanctions and carrying out international criminal prosecutions. Chapter 6 reviewed the key models and theories from the field of regulation that would be later be deployed in the form of the dual networked pyramid regulatory model. The salient features of the dual networked pyramid regulatory model were set out in Chapter 7, with networks theory explained by processes of persuasion and socialisation, while pyramid elements established a coherent rationale for deploying more coercive interventions in appropriate cases. Chapter 8 involved responding to the two original research questions by applying the dual networked pyramid model in descriptive and normative terms to the conflict diamonds governance system.

In response to the question of whether the conflict diamonds governance system has achieved its objectives, I have argued that, with the reduction of conflict diamonds traffic to less than 1 per cent of the world's diamond trade, there has been a marked improvement since the pre-Kimberley era. It is also arguable that the success of peace initiatives in Sierra Leone, Angola, Liberia, and Côte d'Ivoire have been in part attributable to the conflict diamonds governance system. Despite ongoing conflict in the Democratic Republic of Congo (DRC) and Central African Republic, the Kimberley Process has arguably contributed to the emergence of peace and stability in these countries through its handling of blood diamond issues. However, it would appear that it has fallen short of the mark in its response to the emergence of conflict diamonds in Zimbabwe. The seeming inability of the system to grapple with persistent non-compliance by Venezuela has posed a further challenge.

Analysis of the reasons for the conflict diamonds governance system's level of success to date must start with its proven ability to bring together the vast majority of the rough diamond industry, national governments, and concerned NGOs to combat the issue. In an industry that was once considered opaque, the publication of global diamond statistics is a huge breakthrough in transparency. Other monitoring, through annual reports and peer reviews, has a proven ability to highlight relevant issues of non-compliance. The Kimberley Process has even had some level of success with situations of serious non-compliance. On the lower end of the scale, efficacy of the peer-review system to promote normative compliance was demonstrated in the case of Brazil. Once its problems had been highlighted by an NGO report, and confirmed by a Kimberley review visit, Brazil acted conscientiously in taking legal action against illicit trafficking. Another example of success was the initiative taken by the Kimberley Chair, occupied at that time by the Canadian Government, in expelling the Republic of Congo from membership when it became clear that that country was a conduit for smuggled DRC diamonds. The international spotlight in 2017 will be on Australia, as it takes its turn in the difficult position of Chair of the Kimberley Process.

The conflict diamonds governance system has not been an unqualified success. It faces three major problems: the continued role of conflict diamonds in ongoing conflicts in Côte d'Ivoire, the DRC and Zimbabwe; the serious non-compliance and active resistance of two of its members, Zimbabwe and Venezuela; and the threat that, ironically, the Kimberley Process may become a victim of its own success, with some governments calling for it to be dissolved. Of these threats, the inability of the Kimberley Process to deal with the active resistance of its own membership is the most serious. This threat represents not uncertainty in the potential result of Kimberley Process action, but actual failure to take action. While the continued flow of Venezuela's diamonds represents an open breach by which conflict diamonds might contaminate the Kimberley chain of custody, the Zimbabwean case is worse in that it enables the continued flow of conflict diamonds into the Kimberley system. The Zimbabwe case reflects a number of systemic challenges within the KP itself. Most significant is the inability of the KP to take a purposive approach to understanding the definition of conflict diamonds so that it might confront human rights abuses committed by governments and the perpetuation of conflict through fuelling the financial coffers of militia groups. Alternatively, or in parallel, the KP might have sought an amendment of its definition to cover these circumstances. A related issue concerns the definition of consensus under which the organisation operated. With consensus understood as unanimity, the ability of the organisation to move forward decisively has been hampered. No doubt the lack of a permanent secretariat and budget constraints have also hindered decisive intervention in the face of non-compliance.

It appears a little ironic that the Kimberley Process was able to deal with serious non-compliance by the Republic of Congo early in its mandate, with the chair expelling it from membership following a rapidly deployed review visit, but similar action has not been forthcoming in the cases of Zimbabwe and Venezuela. Certainly, this type of discrepancy leaves the Kimberley Process open to the criticism, as suggested by Global Witness in its research interview response, that the effectiveness of the response is largely dictated by which person is the occupant of the Kimberley Process rotating chair. Perhaps the same argument, put differently, is that there is no clear procedure for dealing with situations of serious non-compliance, including which body (whether the chair or the plenary) is empowered to act. If it is the plenary which is so empowered, there is a further problem in relation to the choosing of consensus as the mode of decision making. Consensus by the plenary voting membership in a decision to expel a national government from the Kimberley Process for serious non-compliance is well-nigh impossible to achieve. If expulsion is to be a regulatory ratchet available to the Kimberley Process, then a majority vote by the plenary is the only viable option.

The Kimberley Process has proven itself able to evolve since the decision to finalise its founding agreement in 2002. Two important examples of this are the peer-review mechanism and the Artisanal Diamond Working Group. When the Kimberley Process Agreement was finalised, there was no monitoring mechanism included in its provisions. Led by NGOs, the Kimberley Process agreed within its own plenary membership to the creation of the Working Group on Monitoring, the Participation Committee, and the central mechanism of peer review, which has been central to the amount of success that the Kimberley Process has enjoyed to date. Another example, also on the initiative of NGOs, was the Artisanal Mining Working Group. Aware of the centrality of artisanal diamond mining to the problem of conflict diamonds, the plenary was able to approve the creation of a committee mandated to assist with the working conditions of artisanal miners globally. It is this ability to evolve that is now being challenged in relation to its ability to deal with issues of serious non-compliance. Also needing to be addressed is formalisation within the Kimberley Process Agreement, backed up by UNSC resolution, that the definition of conflict diamonds includes diamonds that are connected to human rights violations even in situations where there is no ongoing conflict.

Armed with a clear picture of the success and failure of the conflict diamonds governance system, it is possible to consider insights gained by analysing it in the light of the networked pyramid regulatory model. The coming together of NGOs, industry, and governments to create the Kimberley Process is almost by definition a network approach. It has almost exclusively relied upon methods of persuasion and socialisation, rather than coercion and punitive action. The discussion that occurs in the Kimberley Process Plenary, the information shared by industry about technical issues, or NGOs about compliance issues, and the informal naming and shaming as a result of peer-review visits all represent the standard tool-kit of networked governance. Such approaches confirm what has already been suggested in the literature, that networked governance can achieve a great deal. However, the Kimberley Process has a contribution to make towards networks theory beyond this statement. One of the features of the Kimberley Process that is not accounted for in networks theory is that fact that, as a network, it embodies a separation of interests. This constitutionalist model operates to create checks and balances in the regulatory operation of the Kimberley Process. As such, it represents a very different type of network to the networked regulation of the intellectual property regime. While the initial driver for the intellectual property regime was the profit motive of corporations, supported by the US national government, the Kimberley Process set itself up, at least in the short term, as a break on free-for-all profit maximisation by the international diamond industry. The role of NGOs and enlightened national governments has been to help socialise big business into choosing long-term, enlightened self-interest over short-term profit maximisation. It has been remarkably successful in this effort, with the diamond industry major players coming on board. This task has naturally been a lot more difficult than promoting immediate self-interest. The separation of interests model establishes a more robust organisational model whereby such a socialisation process can occur. It now faces significant challenges in confronting the immediate self-interest of a number of key national government participants.

The networks theory concept of a node is a valuable tool in recognising the potential of the Kimberley Process to ratchet up interventions by engaging with higher level regulators, such as the UNSC and the international criminal tribunals. The further insights into enhancing the conflict diamonds governance system, however, relate more to the domain of pyramid than networks theory. Part of the success of the Kimberley Process has to be credited to the ongoing efforts of the UNSC and the international tribunals. However, there are a number of ways in which the regulatory ratchet from expulsion to UN sanctions to international prosecutions can be strengthened. Chief amongst these is the creation of an international crime of trafficking in conflict diamonds, which is defined in the ICC statute in terms of a contravention of UNSC diamond-trading sanctions. With the benefit of such an amendment, the threat of escalation to the UNSC and then the ICC becomes more meaningful, thereby enhancing the ability of the Kimberley Process to carry a big stick at the same time as it speaks softly. The enactment of a crime of trafficking in conflict diamonds at the domestic level would be a logical progression of this connection. It would create a fitting apex to the regulatory pyramid at the domestic level: the pyramid within the pyramid. The base of the national pyramid also needs attention, as national government backing to initiatives such as the Diamond Development Initiative International would assist working conditions for artisanal diamond miners, thereby reducing the incentive to sell to illegal operatives engaged in human rights abuses.

Further Research

Further research projects might relate to the theoretical and empirical investigations of this book. On the theoretical side, the dual networked pyramid model might usefully be applied to other complex international systems, such as the intellectual property system, so as to determine its strengths and weaknesses in other contexts. In relation to the conflict diamonds governance system, further studies might investigate the extent to which the KP is responsive to undertaking a wider mandate based around the concept of development diamonds.

Concluding Remarks

This chapter has summarised the findings of the book in response to the two research questions. The first question sought to assess how effective the conflict diamonds governance system has been in meeting its goals. In response, the book argues that it has achieved a measure of success, notably reducing the quantity of conflict diamonds in the international system from estimates as high as 15 per cent in the 1990s to less than 1 per cent in recent years,[2] and establishing an innovative tripartite partnership that has socialised the major diamond industry players into becoming active proponents of the Kimberley Process. The system as a whole is able to draw on a range of enforcement mechanisms, from horizontal peer review and informal naming and shaming, to expulsion for serious non-compliance, and, through its networked relationship with the UNSC and international tribunals, UNSC diamond embargoes and international criminal prosecutions.

Despite the measure of success it has achieved, the Kimberley Process and its collaborators have fallen into a state of deepening crisis, beginning with its inability to manage serious non-compliance by Venezuela in 2006. Even worse, the KP did not act decisively following the commission of international human rights crimes by Zimbabwe and Angola in artisanal diamond fields in those countries. This inability to quarantine conflict diamonds from Zimbabwe and Angola from the KP regime continues to undermine the legitimacy of the conflict diamonds system and may lead to complete system failure.

---

2   For discussion about estimates of the quantity of the conflict diamonds trade, see Chapter 2.

The second question asks whether the dual networked pyramid model offers a description of how conflict diamonds governance currently operates or if it may be deployed normatively so as to suggest ways in which the system might be improved. The networked pyramid model is a hybrid, combining insights from networks theory, the regulatory pyramid model, and the pyramid of rewards. Networks theory suggests that regulation occurs through the combined operation of different individuals and organisations, which are considered roughly on equal terms. Its main regulatory techniques are dialogue and persuasion, and informal naming and shaming, which together create a process of socialisation towards compliance with a particular set of standards. Networks theory is an intuitive fit for the Kimberley Process, which combines networks of national governments, NGOs and diamond industry corporations. Acting as a command centre or node for these networks, the Kimberley Process is able to collect information and deploy a range of regulatory interventions, including peer-review reporting, which largely resembles the informal naming and shaming of networks theory. Indeed, one of the most dramatic examples of socialisation in the KP is the manner in which major diamond corporation, and alleged conflict diamonds trader, De Beers, became a stalwart proponent of the new system.

It is, however, the regulatory pyramid model that is best able to give insight into the deployment of more coercive interventions, where these are appropriate. The conflict diamonds governance system intuitively fits this model, with regulation at the national level regulated by the informal naming and shaming through the KP peer-review process. The system appears to be floundering, however, in being capable of ratchetting up from here to expulsion from the KP in the event of serious non-compliance. To achieve this, the KP needs to clearly define what constitutes serious non-compliance, clarify that the definition of conflict diamonds includes diamonds connected to grave human rights abuses, abandon the consensus approach to these issues when considered by the plenary, and empower the rotating Kimberley Process Chair to take expeditious expulsion decisions where time is of the essence.

Further improvements to current KP governance involve strengthening links to regulators at higher levels of the regulatory pyramid, namely the UNSC and the ICC. The UNSC has at its disposal a more powerful diamond embargo than the KP, which is binding under international law. It would serve to strengthen the effectiveness of the system if the UNSC saw itself as a type of appeals body in relation to the KP. That way, a serious non-compliance issue might be forwarded to it for consideration, or it might consider the issue on its own motion, and impose a diamond embargo in situations where the KP has failed to act for political or other reasons. Such action would strengthen the hand of the KP, making it clear to member states in serious non-compliance that the UNSC will take action if the KP doesn't. Contemplation of conflict diamonds governance in terms of the regulatory pyramid gives further insights into how the ratchetting-up pathway might be strengthened. Legislating for a specific crime of trading in conflict diamonds under the statute of the ICC would be a strengthening measure, particularly if the new crime was defined in terms of breaching UNSC diamond-trading embargoes. If such a crime were created, then the UNSC diamond-trading embargo would be strengthened as an intervention, because sanctions busting would carry with it the possibility of international criminal prosecution. All of these developments at the UNSC and international tribunal levels would strengthen the hand of the KP in carrying out its activities. This is because responsiveness to the KP speaking softly would be increased so as to avoid the possibility of the big stick.

The KP is in a state of crisis and needs to consider seriously the measures suggested above so as to protect the integrity of its activities from the taint of conflict diamonds originating from Zimbabwe and Angola. Without diminishing the importance of this primary activity, an organisation is sometimes bolstered by envisioning a future for itself above and beyond the crisis in which it finds itself. If the KP is to move beyond simply fixing its current problems, and look at a longer-term preventative mandate, it would do well to reflect on insights available from the pyramid of rewards. This model is the carrot to the regulatory pyramid's stick. It promotes the concept of development diamonds voluntarily certified against aspirational standards as a counterpoint to diamonds that must be certified as conflict-free prior to export. If artisanal diamonds are, for example, mined without the use of child labour, they would qualify for the development diamonds label, which would enable them to access greater profits through being sold as a fair trade commodity. Should the KP become the central administrator of this system, it could generate a standards-raising regulatory ratchet, whereby today's aspirational standard becomes tomorrow's mandatory standard, without which rough diamond export would be prohibited. Beyond the certification of development diamonds, a system of escalating rewards might include the practice of naming and faming NGOs, industry groups, and national governments who have made the greatest progress towards aspirational standards. A notch up, grants might be made available for projects supporting the achievement of aspirational standards. At the apex of the pyramid, if it had been merited, might be nomination for a Nobel Peace Prize.

This chapter has also included a number of recommendations to national governments, the Kimberley Process, the UNSC, and international criminal tribunals for action to strengthen the conflict diamonds governance system. Finally, the chapter has suggested new areas for further research, in relation to both theoretical areas of interest and the conflict diamonds governance system.

Appendix: The Kimberley Process Core Document

KIMBERLEY PROCESS CERTIFICATION SCHEME

PREAMBLE

PARTICIPANTS,

RECOGNISING that the trade in conflict diamonds is a matter of serious international concern, which can be directly linked to the fuelling of armed conflict, the activities of rebel movements aimed at undermining or overthrowing legitimate governments, and the illicit traffic in, and proliferation of, armaments, especially small arms and light weapons;

FURTHER RECOGNISING the devastating impact of conflicts fuelled by the trade in conflict diamonds on the peace, safety and security of people in affected countries and the systematic and gross human rights violations that have been perpetrated in such conflicts;

NOTING the negative impact of such conflicts on regional stability and the obligations placed upon states by the United Nations Charter regarding the maintenance of international peace and security;

BEARING IN MIND that urgent international action is imperative to prevent the problem of conflict diamonds from negatively affecting the trade in legitimate diamonds, which makes a critical contribution to the economies of many of the producing, processing, exporting and importing states, especially developing states;

RECALLING all of the relevant resolutions of the United Nations Security Council under Chapter VII of the United Nations Charter, including the relevant provisions of Resolutions 1173 (1998), 1295 (2000), 1306 (2000), and 1343 (2001), and determined to contribute to and support the implementation of the measures provided for in these resolutions;

HIGHLIGHTING the United Nations General Assembly Resolution 55/56 (2000) on the role of the trade in conflict diamonds in fuelling armed conflict, which called on the international community to give urgent and careful consideration to devising effective and pragmatic measures to address this problem;

FURTHER HIGHLIGHTING the recommendation in United Nations General Assembly Resolution 55/56 that the international community develop detailed proposals for a simple and workable international certification scheme for rough diamonds based primarily on national certification schemes and on internationally agreed minimum standards;

RECALLING that the Kimberley Process, which was established to find a solution to the international problem of conflict diamonds, was inclusive of concerned stake holders, namely producing, exporting and importing states, the diamond industry and civil society;

CONVINCED that the opportunity for conflict diamonds to play a role in fuelling armed conflict can be seriously reduced by introducing a certification scheme for rough diamonds designed to exclude conflict diamonds from the legitimate trade;

RECALLING that the Kimberley Process considered that an international certification scheme for rough diamonds, based on national laws and practices and meeting internationally agreed minimum standards, will be the most effective system by which the problem of conflict diamonds could be addressed;

ACKNOWLEDGING the important initiatives already taken to address this problem, in particular by the governments of Angola, the Democratic Republic of Congo, Guinea and Sierra Leone and by other key producing, exporting and importing countries, as well as by the diamond industry, in particular by the World Diamond Council, and by civil society;

WELCOMING voluntary self-regulation initiatives announced by the diamond industry and recognising that a system of such voluntary self-regulation contributes to ensuring an effective internal control system of rough diamonds based upon the international certification scheme for rough diamonds;

RECOGNISING that an international certification scheme for rough diamonds will only be credible if all Participants have established internal systems of control designed to eliminate the presence of conflict diamonds in the chain of producing, exporting and importing rough diamonds within their own territories, while taking into account that differences in production methods and trading practices as well as differences in institutional controls thereof may require different approaches to meet minimum standards;

FURTHER RECOGNISING that the international certification scheme for rough diamonds must be consistent with international law governing international trade;

ACKNOWLEDGING that state sovereignty should be fully respected and the principles of equality, mutual benefits and consensus should be adhered to;

RECOMMEND THE FOLLOWING PROVISIONS:

SECTION I

Definitions

For the purposes of the international certification scheme for rough diamonds (hereinafter referred to as 'the Certification Scheme') the following definitions apply:

CONFLICT DIAMONDS means rough diamonds used by rebel movements or their allies to finance conflict aimed at undermining legitimate governments, as described in relevant United Nations Security Council (UNSC) resolutions insofar as they remain in effect, or in other similar UNSC resolutions which may be adopted in the future, and as understood and recognised in United Nations General Assembly (UNGA) Resolution 55/56, or in other similar UNGA resolutions which may be adopted in future;

COUNTRY OF ORIGIN means the country where a shipment of rough diamonds has been mined or extracted;

COUNTRY OF PROVENANCE means the last Participant from where a shipment of rough diamonds was exported, as recorded on import documentation;

DIAMOND means a natural mineral consisting essentially of pure crystallised carbon in the isometric system, with a hardness on the Mohs (scratch) scale of 10, a specific gravity of approximately 3.52 and a refractive index of 2.42;

EXPORT means the physical leaving/taking out of any part of the geographical territory of a Participant;

EXPORTING AUTHORITY means the authority(ies) or body(ies) designated by a Participant from whose territory a shipment of rough diamonds is leaving, and which are authorised to validate the Kimberley Process Certificate;

FREE TRADE ZONE means a part of the territory of a Participant where any goods introduced are generally regarded, insofar as import duties and taxes are concerned, as being outside the customs territory;

IMPORT means the physical entering/bringing into any part of the geographical territory of a Participant;

IMPORTING AUTHORITY means the authority(ies) or body(ies) designated by a Participant into whose territory a shipment of rough diamonds is imported to conduct all import formalities and particularly the verification of accompanying Kimberley Process Certificates;

KIMBERLEY PROCESS CERTIFICATE means a forgery resistant document with a particular format which identifies a shipment of rough diamonds as being in compliance with the requirements of the Certification Scheme;

OBSERVER means a representative of civil society, the diamond industry, international organisations and non-participating governments invited to take part in Plenary meetings; *(Further consultations to be undertaken by the Chair.)*

PARCEL means one or more diamonds that are packed together and that are not individualised;

PARCEL OF MIXED ORIGIN means a parcel that contains rough diamonds from two or more countries of origin, mixed together;

PARTICIPANT means a state or a regional economic integration organisation for which the Certification Scheme is effective; *(Further consultations to be undertaken by the Chair.)*

REGIONAL ECONOMIC INTEGRATION ORGANISATION means an organisation comprised of sovereign states that have transferred competence to that organisation in respect of matters governed by the Certification Scheme;

ROUGH DIAMONDS means diamonds that are unworked or simply sawn, cleaved or bruted and fall under the Relevant Harmonised Commodity Description and Coding System 7102.10, 7102.21 and 7102.31;

SHIPMENT means one or more parcels that are physically imported or exported;

TRANSIT means the physical passage across the territory of a Participant or a non-Participant, with or without transhipment, warehousing or change in mode of transport, when such passage is only a portion of a complete journey beginning and terminating beyond the frontier of the Participant or non-Participant across whose territory a shipment passes;

SECTION II

The Kimberley Process Certificate

Each Participant should ensure that:

a. a Kimberley Process Certificate (hereafter referred to as the Certificate) accompanies each shipment of rough diamonds on export;

b. its processes for issuing Certificates meet the minimum standards of the Kimberley Process as set out in Section IV;

c. Certificates meet the minimum requirements set out in Annex I. As long as these requirements are met, Participants may at their discretion establish additional characteristics for their own Certificates, for example their form, additional data or security elements;

d. it notifies all other Participants through the Chair of the features of its Certificate as specified in Annex I, for purposes of validation.

SECTION III

Undertakings in respect of the international trade in rough diamonds

Each Participant should:

a. with regard to shipments of rough diamonds exported to a Participant, require that each such shipment is accompanied by a duly validated Certificate;

b. with regard to shipments of rough diamonds imported from a Participant:

- require a duly validated Certificate;

- ensure that confirmation of receipt is sent expeditiously to the relevant Exporting Authority. The confirmation should as a minimum refer to the Certificate number, the number of parcels, the carat weight and the details of the importer and exporter;

- require that the original of the Certificate be readily accessible for a period of no less than three years;

c. ensure that no shipment of rough diamonds is imported from or exported to a non-Participant;

d. recognise that Participants through whose territory shipments transit are not required to meet the requirement of paragraphs (a) and (b) above, and of Section II (a) provided that the designated authorities of the Participant through whose territory a shipment passes, ensure that the shipment leaves its territory in an identical state as it entered its territory (i.e. unopened and not tampered with).

SECTION IV

Internal Controls

Undertakings by Participants

Each Participant should:

a. establish a system of internal controls designed to eliminate the presence of conflict diamonds from shipments of rough diamonds imported into and exported from its territory;

b. designate an Importing and an Exporting Authority(ies);

c. ensure that rough diamonds arc imported and exported in tamper resistant containers;

d. as required, amend or enact appropriate laws or regulations to implement and enforce the Certification Scheme and to maintain dissuasive and proportional penalties for transgressions;

e. collect and maintain relevant official production, import and export data, and collate and exchange such data in accordance with the provisions of Section V;

f. when establishing a system of internal controls, take into account, where appropriate, the further options and recommendations for internal controls as elaborated in Annex II.

Principles of Industry Self-Regulation

Participants understand that a voluntary system of industry self-regulation, as referred to in the Preamble of this Document, will provide for a system of warranties underpinned through verification by independent auditors of individual companies and supported by internal penalties set by industry, which will help to facilitate the full traceability of rough diamond transactions by government authorities.

SECTION V

Co-operation and Transparency

Participants should:

a. provide to each other through the Chair information identifying their designated authorities or bodies responsible for implementing the provisions of this Certification Scheme. Each Participant should provide to other Participants through the Chair information, preferably in electronic format, on its relevant laws, regulations, rules, procedures and practices, and update that information as required. This should include a synopsis in English of the essential content of this information;

b. compile and make available to all other Participants through the Chair statistical data in line with the principles set out in Annex III;

c. exchange on a regular basis experiences and other relevant information, including on self-assessment, in order to arrive at the best practice in given circumstances;

d. consider favourably requests from other Participants for assistance to improve the functioning of the Certification Scheme within their territories;

e. inform another Participant through the Chair if it considers that the laws, regulations, rules, procedures or practices of that other Participant do not ensure the absence of conflict diamonds in the exports of that other Participant;

f. cooperate with other Participants to attempt to resolve problems which may arise from unintentional circumstances and which could lead to non-fulfilment of the minimum requirements for the issuance or acceptance of the Certificates, and inform all other Participants of the essence of the problems encountered and of solutions found;

g. encourage, through their relevant authorities, closer co-operation between law enforcement agencies and between customs agencies of Participants.

SECTION VI

Administrative Matters

MEETINGS

1. Participants and Observers are to meet in Plenary annually, and on other occasions as Participants may deem necessary, in order to discuss the effectiveness of the Certification Scheme.

2. Participants should adopt Rules of Procedure for such meetings at the first Plenary meeting.

3. Meetings are to be held in the country where the Chair is located, unless a Participant or an international organisation offers to host a meeting and this offer has been accepted. The host country should facilitate entry formalities for those attending such meetings.

4. At the end of each Plenary meeting, a Chair would be elected to preside over all Plenary meetings, *ad hoc* working groups and other subsidiary bodies, which might be formed until the conclusion of the next annual Plenary meeting.

5. Participants are to reach decisions by consensus. In the event that consensus proves to be impossible, the Chair is to conduct consultations.

ADMINISTRATIVE SUPPORT

6. For the effective administration of the Certification Scheme, administrative support will be necessary. The modalities and functions of that support should be discussed at the first Plenary meeting, following endorsement by the UN General Assembly.

7. Administrative support could include the following functions:

    a. to serve as a channel of communication, information sharing and consultation between the Participants with regard to matters provided for in this Document;

    b. to maintain and make available for the use of all Participants a collection of those laws, regulations, rules, procedures, practices and statistics notified pursuant to Section V;

    c. to prepare documents and provide administrative support for Plenary and working group meetings;

   d. to undertake such additional responsibilities as the Plenary meetings, or any working group delegated by Plenary meetings, may instruct.

PARTICIPATION

8. Participation in the Certification Scheme is open on a global, non-discriminatory basis to all Applicants willing and able to fulfill the requirements of that Scheme.

9. Any applicant wishing to participate in the Certification Scheme should signify its interest by notifying the Chair through diplomatic channels. This notification should include the information set forth in paragraph (a) of Section V and be circulated to all Participants within one month.

10. Participants intend to invite representatives of civil society, the diamond industry, non-participating governments and international organizations to participate in Plenary meetings as Observers.

PARTICIPANT MEASURES

11. Participants are to prepare, and make available to other Participants, in advance of annual Plenary meetings of the Kimberley Process, information as stipulated in paragraph (a) of Section V outlining how the requirements of the Certification Scheme are being implemented within their respective jurisdictions.

12. The agenda of annual Plenary meetings is to include an item where information as stipulated in paragraph (a) of Section V is reviewed and Participants can provide further details of their respective systems at the request of the Plenary.

13. Where further clarification is needed, Participants at Plenary meetings, upon recommendation by the Chair, can identify and decide on additional verification measures to be undertaken. Such measures are to be implemented in accordance with applicable national and international law. These could include, but need not be limited to measures such as;

   a. requesting additional information and clarification from Participants;

   b. review missions by other Participants or their representatives where there are credible indications of significant non-compliance with the Certification Scheme.

14. Review missions are to be conducted in an analytical, expert and impartial manner with the consent of the Participant concerned. The size, composition, terms of reference and time-frame of these missions should be based on the circumstances and be established by the Chair with the consent of the Participant concerned and in consultation with all Participants.

15. A report on the results of compliance verification measures is to be forwarded to the Chair and to the Participant concerned within three weeks of completion of the mission. Any comments from that Participant as well as the report, are to be posted on the restricted access section of an official Certification Scheme website no later than three weeks after the submission of the report to the Participant concerned. Participants and Observers should make every effort to observe strict confidentiality regarding the issue and the discussions relating to any compliance matter.

COMPLIANCE AND DISPUTE PREVENTION

16. In the event that an issue regarding compliance by a Participant or any other issue regarding the implementation of the Certification Scheme arises, any concerned Participant may so inform the Chair, who is to inform all Participants without delay about the said concern and enter into dialogue on how to address it. Participants and Observers should make every effort to observe strict confidentiality regarding the issue and the discussions relating to any compliance matter.

MODIFICATIONS

17. This document may be modified by consensus of the Participants.

18. Modifications may be proposed by any Participant. Such proposals should be sent in writing to the Chair, at least ninety days before the next Plenary meeting, unless otherwise agreed.

19. The Chair is to circulate any proposed modification expeditiously to all Participants and Observers and place it on the agenda of the next annual Plenary meeting.

REVIEW MECHANISM

20. Participants intend that the Certification Scheme should be subject to periodic review, to allow Participants to conduct a thorough analysis of all elements contained in the scheme. The review should

also include consideration of the continuing requirement for such a scheme, in view of the perception of the Participants, and of international organisations, in particular the United Nations, of the continued threat posed at that time by conflict diamonds. The first such review should take place no later than three years after the effective starting date of the Certification Scheme. The review meeting should normally coincide with the annual Plenary meeting, unless otherwise agreed.

THE START OF THE IMPLEMENTATION OF THE SCHEME

21. The Certification Scheme should be established at the Ministerial Meeting on the Kimberley Process Certification Scheme for Rough Diamonds in Interlaken on 5 November 2002.

ANNEX I

Certificates

A. Minimum requirements for Certificates

A Certificate is to meet the following minimum requirements:

- Each Certificate should bear the title 'Kimberley Process Certificate' and the following statement: 'The rough diamonds in this shipment have been handled in accordance with the provisions of the Kimberley Process Certification Scheme for rough diamonds'
- Country of origin for shipment of parcels of unmixed (i.e. from the same) origin
- Certificates may be issued in any language, provided that an English translation is incorporated
- Unique numbering with the Alpha 2 country code, according to ISO 3166-1
- Tamper and forgery resistant
- Date of issuance
- Date of expiry
- Issuing authority
- Identification of exporter and importer
- Carat weight/mass

- Value in US$
- Number of parcels in shipment
- Relevant Harmonised Commodity Description and Coding System
- Validation of Certificate by the Exporting Authority

B. Optional Certificate Elements

A Certificate may include the following optional features:

- Characteristics of a Certificate (for example as to form, additional data or security elements)
- Quality characteristics of the rough diamonds in the shipment

A recommended import confirmation part should have the following elements:

- Country of destination
- Identification of importer
- Carat/weight and value in US$
- Relevant Harmonised Commodity Description and Coding System
- Date of receipt by Importing Authority
- Authentication by Importing Authority

C. Optional Procedures

Rough diamonds may be shipped in transparent security bags.

The unique Certificate number may be replicated on the container.

ANNEX II

Recommendations as provided for in Section IV, paragraph (f)

General Recommendations

1. Participants may appoint an official coordinator(s) to deal with the implementation of the Certification Scheme.
2. Participants may consider the utility of complementing and/or enhancing the collection and publication of the statistics identified in Annex III based on the contents of Kimberley Process Certificates.

3. Participants are encouraged to maintain the information and data required by Section V on a computerised database.

4. Participants are encouraged to transmit and receive electronic messages in order to support the Certification Scheme.

5. Participants that produce diamonds and that have rebel groups suspected of mining diamonds within their territories are encouraged to identify the areas of rebel diamond mining activity and provide this information to all other Participants. This information should be updated on a regular basis.

6. Participants are encouraged to make known the names of individuals or companies convicted of activities relevant to the purposes of the Certification Scheme to all other Participants through the Chair.

7. Participants are encouraged to ensure that all cash purchases of rough diamonds are routed through official banking channels, supported by verifiable documentation.

8. Participants that produce diamonds should analyse their diamond production under the following headings:
   - Characteristics of diamonds produced
   - Actual production

Recommendations for Control over Diamond Mines

9. Participants are encouraged to ensure that all diamond mines are licensed and to allow only those mines so licensed to mine diamonds.

10. Participants are encouraged to ensure that prospecting and mining companies maintain effective security standards to ensure that conflict diamonds do not contaminate legitimate production.

Recommendations for Participants with Small-scale Diamond Mining

11. All artisanal and informal diamond miners should be licensed and only those persons so licensed should be allowed to mine diamonds.

12. Licensing records should contain the following minimum information: name, address, nationality and/or residence status and the area of authorised diamond mining activity.

Recommendations for Rough Diamond Buyers, Sellers and Exporters

13. All diamond buyers, sellers, exporters, agents and courier companies involved in carrying rough diamonds should be registered and licensed by each Participant's relevant authorities.

14. Licensing records should contain the following minimum information: name, address and nationality and/or residence status.

15. All rough diamond buyers, sellers and exporters should be required by law to keep for a period of five years daily buying, selling or exporting records listing the names of buying or selling clients, their license number and the amount and value of diamonds sold, exported or purchased.

16. The information in paragraph 14 above should be entered into a computerised database, to facilitate the presentation of detailed information relating to the activities of individual rough diamond buyers and sellers.

Recommendations for Export Processes

17. An exporter should submit a rough diamond shipment to the relevant Exporting Authority.

18. The Exporting Authority is encouraged, prior to validating a Certificate, to require an exporter to provide a declaration that the rough diamonds being exported are not conflict diamonds.

19. Rough diamonds should be sealed in a tamper proof container together with the Certificate or a duly authenticated copy. The Exporting Authority should then transmit a detailed e-mail message to the relevant Importing Authority containing information on the carat weight, value, country of origin or provenance, importer and the serial number of the Certificate.

20. The Exporting Authority should record all details of rough diamond shipments on a computerised database.

Recommendations for Import Processes

21. The Importing Authority should receive an e-mail message either before or upon arrival of a rough diamond shipment. The message should contain details such as the carat weight, value, country of origin or provenance, exporter and the serial number of the Certificate.

22. The Importing Authority should inspect the shipment of rough diamonds to verify that the seals and the container have not been tampered with and that the export was performed in accordance with the Certification Scheme.

23. The Importing Authority should open and inspect the contents of the shipment to verify the details declared on the Certificate.

24. Where applicable and when requested, the Importing Authority should send the return slip or import confirmation coupon to the relevant Exporting Authority.

25. The Importing Authority should record all details of rough diamond shipments on a computerised database.

Recommendations on Shipments to and from Free Trade Zones

26. Shipments of rough diamonds to and from free trade zones should be processed by the designated authorities.

Annex III

Statistics

Recognising that reliable and comparable data on the production and the international trade in rough diamonds are an essential tool for the effective implementation of the Certification Scheme, and particularly for identifying any irregularities or anomalies which could indicate that conflict diamonds are entering the legitimate trade, Participants strongly support the following principles, taking into account the need to protect commercially sensitive information:

a. to keep and publish within two months of the reference period and in a standardised format, quarterly aggregate statistics on rough diamond exports and imports, as well as the numbers of certificates validated for export, and of imported shipments accompanied by Certificates;

b. to keep and publish statistics on exports and imports, by origin and provenance wherever possible; by carat weight and value; and under the relevant Harmonised Commodity Description and Coding System (HS) classifications 7102.10; 7102.21; 7102.31;

c. to keep and publish on a semi-annual basis and within two months of the reference period statistics on rough diamond production by carat weight and by value. In the event that a Participant is unable to publish these statistics it should notify the Chair immediately;

d. to collect and publish these statistics by relying in the first instance on existing national processes and methodologies;

e. to make these statistics available to an intergovernmental body or to another appropriate mechanism identified by the Participants for (1) compilation and publication on a quarterly basis in respect of exports and imports, and (2) on a semiannual basis in respect of production. These statistics are to be made available for analysis by interested parties and by the Participants, individually or collectively, according to such terms of reference as may be established by the Participants;

f. to consider statistical information pertaining to the international trade in and production of rough diamonds at annual Plenary meetings, with a view to addressing related issues, and to supporting effective implementation of the Certification Scheme.

Bibliography

Articles

Africa Research Bulletin, 'Diamonds: Zimbabwe' (2011) 41(12) February *Africa Research Bulletin: Economic, Financial and Technical Series* 18960.

Anderson, K, 'Squaring the Circle? Reconciling Sovereignty and Global Governance Through Global Government Networks' (2005) 118 *Harvard Law Review* 1255.

Antal, E, 'Lessons from NAFTA: The Role of the North American Commission for Environmental Cooperation in Conciliating Trade and Environment' (2006) 14 *Michigan State Journal of International Law* 167.

Assaf, Farid, 'What will Trigger ASICs Strategies?' (2002) May *Law Society Journal* 60.

Banat, A B, 'Solving the Problem of Conflict Diamonds in Sierra Leone: Proposed Theories and International Legal Requirements for Certification of Origin' (2002) 19 *Arizona Journal of International and Comparative Law* 939.

Black, B, 'Panel: Combating International Corruption through Law and Institutions' (2007) 5 *Santa Clara Journal of International Law* 445.

Book, L, 'Refund Anticipation Loans and the Tax Gap' (2009) 20 *Stanford Law and Policy Review*.

Braithwaite, John, 'Methods of Power for Development: Weapons of the Weak, Weapons of the Strong' (2005) 26 *Michigan Journal of International Law* 297.

Brockman, J, 'Liberia: The Case for Changing UN Processes for Humanitarian Interventions' (2004) 22 *Wisconsin International Law Journal* 711.

Burris, Scott, Peter Drahos and Clifford Shearing, 'Nodal Governance' (2005) 30 *Australian Journal of Legal Philosophy* 30.

Calton, Jerry M and Steven L Payne, 'Coping With Paradox Multistakeholder Learning Dialogue as a Pluralist Sensemaking Process for Addressing Messy Problems' (2003) 42(1) *Business and Society* 7.

Chung, C, 'International Law and the Extraordinary Interaction Between the People's Republic of China and the Republic of China on Taiwan' (2009) 19 *Indiana International and Comparative Law Review* 233.

Coen, D and M Thatcher, 'Network Governance and Multi-level Delegation: European Networks of Regulatory Agencies' (2008) 28(1) *Journal of Public Policy* 49.

Crane, D M, 'Terrorists, Warlords and Thugs' (2006) 21 *American University International Law Review* 505.

Cuellar, M, 'Panel: Combating Diamonds in Sub-Saharan Africa' (2007) 1 *African Journal of Legal Studies* 80.

Curtis, K, 'But is it Law?: An Analysis on the Legal Nature of the Kimberley Process Certification Scheme on Conflict Diamonds and its Treatment of Non-State Actors' (2007) Spring *The American University International Law Review*.

Drahos, P, 'Intellectual Property and Pharmaceutical Markets: A Nodal Governance Approach' (2004) 77 Summer *Temple Law Review* 401.

Fishman, J L, 'Is Diamond Smuggling Forever?: The Kimberley Process Certification Scheme: The First Step Down the Long Road to Solving the Blood Diamond Trade Problem' (2005) 13 *University of Miami Business Law Review* 217.

Golden Gate University School of Law, 'Annex: Arrest Warrant of 11 April 2000 (Democratic Republic of the Congo v. Belgium): International Court of Justice 14 February 2002' (2002) 8 *Annual Survey of International and Comparative Law* 151.

Gooch, C, 'Global Witness Founding Director's Statement on NGO Coalition Walk-Out from Kimberley Process Meeting', 27 June 2011. Available at: www.globalwitness.org/en/archive/global-witness-founding-directors-statement-ngo-coalition-walk-out-kimberley-process/.

Grabosky, P N, 'Discussion Paper: Inside the Pyramid: Conceptual Framework for the Analysis of Regulatory Systems' (1997) 25 *International Journal of the Sociology of Laws* 195.

Grant, A J and I Taylor, 'Global Governance and Conflict Diamonds: The Kimberley Process and the Quest for Clean Gems' (2004) 93(375) *The Round Table* 385.

Gray, C, 'The Use and Abuse of the International Court of Justice: Cases Concerning the Use of Force After Nicaragua' (2003) 14 *European Journal of International Law* 867.

Hewitt, Tara, 'Implementation and Enforcement of the Convention on International Trade in Endangered Species of Wild Fauna and Flora in the South Pacific Region: Management and Scientific Authorities' (2002) 2(1) *Queensland University of Technology Law Journal* 98.

Holmes, J, 'The Kimberley Process: Evidence of Change in International Law' (2007) 3 *Brigham Young University International Law and Management Review* 213.

Howse, R, 'Book Review: A New World Order, By Anne-Marie Slaughter, Princeton, NJ: Princeton University Press, 2004' (2007) 101 *American Journal of International Law* 231.

Jalloh, C C, 'The Contribution of the Special Court for Sierra Leone to the Development of International Law' (2007) 15 *RADIC* 165.

Juma, L, 'Africa, its Conflicts and its Traditions: Debating a Suitable Role for Tradition in African Peace Initiatives' (2005) 13 *Michigan State Journal of International Law* 417.

Juma, L, 'The War in Congo: Transitional Conflict Networks and the Failure of Internationalism' (2006) 10 *Gonzaga Journal of International Law* 97.

Kantz, C, 'The Power of Socialization: Engaging the Diamond Industry in the Kimberley Process' (2007) 9(3) *Business and Politics* Art 2.

Kaplan, M, 'Junior Fellows' Note: Carats and Sticks: Pursuing War and Peace through the Diamond Trade' (2003) 35 *New York University Journal of International Law and Politics* 559.

Kofele-Kale, N, 'The Global Community's Role in Promoting the Right to Democratic Governance and Free Choice in the Third World' (2005) 11 *Law and Business Review of the Americas* 205.

Koyame, M, 'United Nations Resolutions and the Struggle to Curb the Illicit Trade in Conflict Diamonds in Sub-Saharan Africa' (2005) 1 *African Journal of Legal Studies* 80.

Lang, A and J Scott, 'The Hidden World of WTO Governance' (2009) 20 *European Journal of International* Law 575.

Maggi, M, 'The Currency of Terrorism: An Alternative Way to Combat Terrorism and End the Trade of Conflict Diamonds' (2003) 15 *Pace International Law Review* 513.

Malamut, S A, 'A Band-Aid on a Machete Wound: The Failures of the Kimberley Process and Diamond-Caused Bloodshed in the Democratic Republic of Congo' (2005) 29 *Suffolk Transnational Law Review* 25.

Martinez, I, 'Africa at the Crossroads: Current Themes in African Law: VI. Conflict Resolution in Africa: Sierra Leone's "Conflict Diamonds": The Legacy of Imperial Mining Laws and Policy' (2001/2002) 10 *University of Miami International and Comparative Law Review* 217.

Mena, S and G. Palazzo, 'Input and Output Legitimacy of Multi-stakeholder Initiatives' (2012) 22(3) *Business Ethics Quarterly* 527.

Mikler, J, 'Sharing Sovereignty for Global Regulation: The Cases of Fuel Economy and Online Gambling' (2008) 2(4) *Regulation and Governance* 383.

Mitchell III, A F, 'Sierra Leone: The Road to Childhood Ruination Through Forced Recruitment of Child Soldiers and the World's Failure to Act' (2003) 2 *Regent Journal of International Law* 81.

Montague, D and F Berrigan, 'The Business of War in the Democratic Republic of Congo', *Dollars and Sense Magazine*, July/August 2001. Available at: www.thirdworldtraveler.com/Africa/Business_War_Congo.html.

O'Rourke, D, 'Multi-stakeholder Regulation: Privatizing or Socializing Global Labour Standards?' (2006) 34(5) *World Development*.

Parker, C, 'The "Compliance" Trap: The Moral Message in Responsive Regulatory Enforcement' (2006) 40 *Law and Society Review* 591.

Partnership Africa Canada, 'NGOs Walkout of Kinshasa KP Meeting, Consider Options', *Other Facets: News and Views on the International Effort to End Conflict Diamonds*, (No. 35, August 2011), p. 1. Available at: www.pacweb.org/images/PUBLICATIONS/Other_Facets/OF35-eng.pdf.

Petrova, P, 'The Implementation and Effectiveness of the Kimberley Process Certification Scheme in the United States' (2006) 40 *International Lawyer* 945.

Pham, J P, 'A Viable Model for International Criminal Justice: The Special Court for Sierra Leone' (2006) 19 *New York International Law Review* 37.

Price, T M, 'Article: The Kimberley Process: Conflict Diamonds, WTO Obligations, and the Universality Debate' (2003) 12 *Minnesota Journal of Global Trade* 1.

Punyasena, W, 'Conflict Prevention and the International Criminal Court: Deterrence in a Changing World' (2006) 14 *Michigan State Journal of International Law* 39.

Rawlings, Gregory, 'Taxes and Transitional Treaties: Responsive Regulation and the Reassertion of Offshore Sovereignty' (2007) 27(1) *Law and Policy* 51.

Ross, M, 'How Do Natural Resources Influence Civil War?: Evidence from Thirteen Cases' (2004) 58 Winter *International Organization* 35.

Salo, Rudy S, 'When the Logs Roll Over: The Need for an International Convention Criminalizing Involvement in the Global Illegal Timber Trade' (2003) 16 *Georgetown International Environmental Law Review* 127.

Saunders, L, 'Note: Rich and Rare are the Gems they War: Holding De Beers Accountable for Trading Conflict Diamonds' (2001) 24 *Fordham International Law Journal* 1402.

Schocken, C, 'The Special Court for Sierra Leone: Overview and Recommendations' (2002) 20 *Berkeley Journal of International Law* 436.

Shaw, Timothy M, 'Regional Dimensions of Conflict and Peace-Building in Contemporary Africa' (2003) 15 *Journal of International Development* 487.

Tailby, R, 'The Illicit Market in Diamonds' (2002) 218 *Australian Institute of Criminology: Trends and Issues in Crime and Criminal Justice* 1.

Tamm, Ingrid J, 'Dangerous Appetites: Human Rights Activism and Conflict Commodities' (2004) 26 *Human Rights Quarterly* 687.

Taylor, Ian and Gladys Mkhawa, 'Not Forever: Botswana, Conflict Diamonds and the Bushmen' (2003) 102 *African Affairs* 261.

Torbey, C, 'The Most Egregious Arms Broker: Prosecuting Arms Embargo Violators in the International Criminal Court' (2007) 25 *Wisconsin International Law Journal* 335.

Tripathi, Sali, 'International Regulation of Multinational Corporations' (2005) 33(1) *Oxford Development Studies* 117.

United Kingdom Wildlife Licensing and Registration Service, 'Guidance for Antique Dealers on the Control of Trade in Endangered Species' (2005). Available at: www.culturecommunication.gouv.fr/content/download/97744/875972/version/1/file/2005_Guidance+for+Antique+dealers+on+the+control+of+trade+in+endangered+species.pdf.

Wallis, A, 'Data Mining: Lessons from the Kimberley Process for the United Nations' Development of Human Rights Norms for Transnational Corporations' (2005) 4(2) *Northwestern Journal of International Human Rights* 388.

Williams, C A, 'Civil Society Initiatives and 'Soft Law' in the Oil and Gas Industry' (2004) 36 Winter–Spring *New York University Journal of International Law and Politics* 457.

Woodward, L, 'Taylor's Liberia and the UN's Involvement' (2003) 19 *New York Law School Journal of Human Rights* 923.

Woody, K E, 'Diamonds on the Soul of Her Shoes: The Kimberley Process and the Morality Exception to WTO Restrictions' (2007) 22 *Connecticut Journal of International Law* 335.

Wright, Clive, 'Tackling Conflict Diamonds: The Kimberley Process Certification Scheme' (2004) 11(4) *International Peacekeeping* 697.

Books

Abbott, Kenneth and Duncan Snidal 'The Governance Triangle' in Walter Mattli and Ngairi N Woods (eds), *The Politics of Global Regulation* (Princeton University Press, 2009).

Ayres, Ian and John Braithwaite, *Responsive Regulation: Transcending the Deregulation Debate* (Oxford University Press, 1992).

Bahá'u'lláh, *Tablets of Bahá'u'lláh Revealed After the Kitáb-i-Aqdas* (Bahá'í Publishing Trust 1988).

Beah, Ishmael, *A Long Way Gone: Memoirs of a Boy Soldier* (Sarah Chrichton Books, 2007).

Bieri, Franziska, *From Blood Diamonds to the Kimberley Process: How NGOs Cleaned Up the Diamond Industry* (Ashgate Publishing Ltd, 2010).

Braithwaite, John, *Regulatory Capitalism* (Edward Elgar, 2008).

Braithwaite, John and Peter Drahos, *Global Business Regulation* (Cambridge University Press, 2000).

Campbell, Greg, *Blood Diamonds: Tracing the Deadly Path of the World's Most Precious Stones* (Basic Books, 2013).

Canan, Penelope and Nancy Reichman, *Ozone Connections: Expert Networks in Global Environment Governance* (Greenleaf Publishing Limited, 2002).

Chayes, A and A H Chayes, *The New Sovereignty: Compliance with International Regulatory Agreements* (Harvard University Press, 1995).

Clark, J F and M Koyame, 'The Economic Impact of the Congo War' in J F Clark (ed.), *The Africa Stakes of the Congo War* (Palgrave MacMillan, 2002).

Collins, Hugh, 'Regulating Contract Law' in Christine Parker, et al. (eds), *Regulating Law* (Oxford University Press, 2004).

Dallaire, R, *Shake Hands with the Devil: The Failure of Humanity in Rwanda* (Random House Canada, 2003).

Dunn, K C, 'Identity, Space and the Political Economy of Conflict in Central Africa' in P Le Billon (ed.), *The Geopolitics of Resource Wars: Resource Dependence, Governance and Violence* (Frank Cass, 2005).

Freiberg, Arie, *The Tools of Regulation* (The Federation Press, 2010).

Koechlin, Lucy and Richard Calland, 'Standard-setting at the Cutting Edge: An Evidence-based Typology for Multi-stakeholder Initiatives' in Anne Peters et al. (eds), *Non-State Actors as Standard Setters* (Cambridge University Press, 2009).

Le Billon, Philippe, 'The Geopolitical Economy of "Resource Wars"' in P Le Billon (ed.), *The Geopolitics of Resource Wars: Resource Dependence, Governance and Violence* (Frank Cass, 2005).

Parker, Christine et al. (eds), *Regulating Law* (Oxford University Press, 2004).

Salamon, L M, 'The New Governance and the Tools of Public Action: An Introduction' in L M Salamon (ed.), *The Tools of Government: A Guide to the New Governance* (Oxford University Press, 2002).

Scott, Colin, 'Regulation in the Age of Governance: The Rise of the Post-regulatory State' in Jacint Jordana and David Levi-Fuar (eds), *The Politics of Regulation: Institutions and Regulatory Reform for the Age of Governance* (Edward Elgar, 2004).

Slaughter, Anne-Marie, *A New World Order* (Princeton University Press, 2004).

Smillie, Ian, *Blood on the Stone: Greed, Corruption and War in the Global Diamond Trade* (Anthem Press, 2010).

Sprecher, Drexel A, *Inside the Nuremberg Trial: A Prosecutor's Comprehensive Account*, Vol. 1 (University Press of America, 1999).

Werner, J R and J Zimmermann, 'The Evolving Post-National Regulation of Financial Reporting' in H Rothgang and S Schneider (eds), *State Transformations in OECD Countries: Dimensions, Driving Forces, and Trajectories* (Palgrave Macmillan, 2015).

Witte, J M, T Benner and C Streck 'Partnerships and networks in global environmental governance', in U. Petschow, J. Rosenau and E U von Weizsacker (eds), *Governance and Sustainability: New Challenges for States, Companies and Civil Society* (Greenleaf, 2005).

Cases

Case Concerning Armed Activities on the Territory of the Congo (Democratic Republic of Congo v Uganda) (Judgement) (1995) ICJ Rep 90.

Doe I v Unocal Corp, 248 F 3d 915, (9th Cir, 2001).

Guus Kouwenhoven, Rechtbank 's-Gravenhage [District Court of The Hague], Case No AY5160, 7 June 2006.

Guus Kouwenhoven, Rechtbank 's-Gravenhage [The Hague Court of Appeal], Case No BC7373, 10 March 2008.

The Prosecutor v Alex Tamba Brima, Brima Bazzy Kamara and Santigie Borbor Kanu (Trial Judgement) (Special Court for Sierra Leone, Trial Chamber II, Case No SCSL-04-16-T, 20 June 2007).

The Prosecutor v Alex Tamba Brima, Brima Bazzy Kamara and Santigie Borbor Kanu (Appeal Judgement) (Special Court for Sierra Leone, Appeals Chamber, Case No SCSL-2004-16-A, 22 February 2008).

The Prosecutor v Bosco Ntaganda (Decision on Prosecutor's Application for Warrants of Arrest, Article 58) (International Criminal Court, Pre-Trial Chamber I, Case No ICC-01/04-02/06, 10 February 2006).

The Prosecutor v Brdjanin (Appeal Judgement) (International Criminal Tribunal for the Former Yugoslavia, Appeals Chamber, Case No IT-99-36-A, 3 April 2007).

The Prosecutor v Dusko Tadic (Appeal Judgement) (International Criminal Tribunal for the Former Yugoslavia, Appeals Chamber, Case No IT-94-1-A, 15 July 1999).

The Prosecutor v Issa Hassan Sesay, Morris Kallon and Augustine Gbao (Trial Judgement) (Special Court for Sierra Leone, Trial Chamber I, Case No SCSL-04-15-T, 2 March 2009).

The Prosecutor v Issa Hassan Sesay, Morris Kallon and Augustine Gbao (Appeal Judgement) (Special Court for Sierra Leone, Appeals Chamber, 29 October 2009).

The Prosecutor v Moinina Fofana and Allieu Kondewa (Trial Judgement) (Special Court for Sierra Leone, Trial Chamber I, Case No SCSL-04-14-T, 2 August 2007).

The Prosecutor v Moinina Fofana and Allieu Kondewa (Appeal Judgement) (Special Court for Sierra Leone, Appeals Chamber, Case No SCSL-04-14-A, 28 May 2008).

Residual Special Court for Sierra Leone. Available at: www.rscsl.org.

Trials of Nazi War Criminals before the Neurnberg Tribunals under Control Council Law No 10: Neurnberg October 1946–April 1949, Volumes V, VI, VII, VIII, IX (United States Government Printing Office, 1952). Available at: www.phdn.org/archives/www.mazal.org/NMT-HOME.htm.

Case Documents

'Corrected Amended Consolidated Indictment', *The Prosecutor v Issa Hassan Sesa, Morris Kallon and Augustine Gbao* (Special Court for Sierra Leone, Case No SCSL-2004-15-PT, 2 August 2006).

'Document Containing the Charges Pursuant to Article 61(3)(a) of the Statute', *The Prosecutor v Germain Katanga and Mathieu Ngudjolo Chui* (International Criminal Court, Case No ICC-01/04-01/07, 21 April 2008).

'Further Amended Consolidated Indictment', *The Prosecutor v Alex Tamba Brima, Brima Bazzy Kamara and Santigie Borbor Kanu* (Special Court for Sierra Leone, Case No SCSL-2004-16-PT, 18 February 2005).

'Indictment', *The Prosecutor v Samuel Hinga Norman, Moinina Fofana and Allieu Kondewa* (Special Court for Sierra Leone, Case No SCSL-03-14-I, 5 February 2004).

International Criminal Court, 'Case Information Sheet: The Prosecutor v. Germain Katanga', ICC-01/04-01/07. Available at: www.icc-cpi.int/iccdocs/PIDS/publications/KatangaEng.pdf.

International Criminal Court, 'Case Information Sheet: The Prosecutor v. Mathieu Ngudjolo Chui', ICC-01/04-02/12. Available at: www.icc-cpi.int/iccdocs/PIDS/publications/ChuiEng.pdf.

International Criminal Court, 'Case Information Sheet: The Prosecutor v. Thomas Lubanga Dyilo', ICC-01/04-01/06. Available at: www.icc-cpi.int/iccdocs/PIDS/publications/LubangaENG.pdf.

International Criminal Court, Office of the Prosecutor, 'First Arrest for the International Criminal Court' (Press Release, ICC-CPI-20060302-125-En, 17 March 2006).

International Criminal Court, Office of the Prosecutor, 'Issuance of a Warrant of Arrest against Thomas Lubanga' (Press Release, ICC-OTP-20060302-126-En, 2 March 2006).

International Criminal Court, Office of the Prosecutor, *The Prosecutor on the Co-operation with Congo and other States Regarding the Situation in Ituri, DRC* (26 September 2003).

The International Criminal Court, *The Prosecutor vs. Laurent Gbagbo and Ble Goude,* ICC-02/11-01/15. Available at: www.icc-cpi.int/cdi/gbagbo-goude/Pages/default.aspx.

International Criminal Court, 'Case Information Sheet for The Prosecutor v. Germain Katanga, ICC-01/04-01/07'. Available at: www.icc-cpi.int/en_menus/icc/situations%20and%20cases/situations/situation%20icc%200104/related%20cases/icc%200104%200107/Pages/democratic%20republic%20of%20the%20congo.aspx.

International Criminal Court, 'Case Information Sheet for The Prosecutor v. Mathieu Ngudjolo Chui, ICC-01/04-02/12', www.icc-cpi.int/en_menus/icc/situations%20and%20cases/situations/situation%20icc%200104/related%20cases/ICC-01-04-02-12/Pages/default.aspx.

International Criminal Court, Pre-Trial Chamber III, 'Decision Pursuant to Article 15 of the Rome Statute on the Authorisation of an Investigation into the Situation in the Republic of Côte d'Ivoire' (Pre-Trial Chamber III Decision, ICC-02/11, 3 October 2011) paras 10–15, 34–35, 212–213. Available at: www.icc-cpi.int/Pages/record.aspx?docNo=ICC-02/11-14&ln=en.

'Prosecution's Second Amended Indictment', *The Prosecutor v Charles Taylor* (Special Court for Sierra Leone, Case No SCSL-03-01-PT, 29 May 2007).

Special Court for Sierra Leone, Office of the Prosecutor, 'Chief Prosecutor Announces the Arrival of Charles Taylor at the Special Court' (Press Release, 29 March 2006). Available at:www.rscsl.org/Documents/Press/OTP/prosecutor-032906.pdf.

The Special Court for Sierra Leone and the Residual Court for Sierra Leone at Freetown and The Hague, *The Prosecutor vs. Charles Gankay Taylor*. Available at: www.rscsl.org/Taylor.html.

Transcript of Proceedings, 'Evidence of Expert Witness Ian Smillie', *The Prosecutor v Charles Taylor (Trial)*, (Special Court for Sierra Leone, Trial Chamber II, Case No SCSL-03-01-PT, 7 January 2008) 550–562.

Transcript of Proceedings, 'Opening Statement of the Prosecution', *The Prosecutor v Alex Tamba Brima, Brima Bazzy Kamara and Santigie Borbor Kanu (Trial)*, (Special Court for Sierra Leone, Trial Chamber I, Case No SCSL-2004-16-T, 7 March 2005).

Transcript of Proceedings, 'Opening Statement of the Prosecution', *The Prosecutor v Issa Hassan Sesay, Morris Kallon and Augustine Gbao (Trial)*, (Trial Chamber I, Special Court for Sierra Leone, Case No SCSL-04-15-T, 5 July 2004).

Transcript of Proceedings, 'Opening Statement of the Prosecution', *The Prosecutor v Moinina Fofana and Allieu Kondewa (Trial)*, (Special Court for Sierra Leone, Trial Chamber I, Case No SCSL-04-14-T, 3 June 2003).

Transcript of Proceedings, 'Prosecutor's Opening Statement', *The Prosecutor v Thomas Lubanga Dyilo (Trial)*, (International Criminal Court, Trial Chamber I, Case No ICC-01/04-01/06, 26 January 2009).

Transcript of Proceedings, *The Prosecutor v Charles Taylor (Trial)*, (Special Court for Sierra Leone, Trial Chamber II, Case No SCSL-03-01-PT, 4 June 2007).

'Warrant of Arrest', *The Prosecutor v Bosco Ntaganda* (International Criminal Court, Case No ICC-01/04-02/06, 7 August 2006).

'Warrant of Arrest', *The Prosecutor v Thomas Lubanga Dyilo* (International Criminal Court, Case No ICC-01/04-01/06, 10 February 2006).

Kimberley Process: Miscellaneous Documents

Ad Hoc Working Group on the Review of the Kimberley Process Certification Scheme, *Kimberley Process Certification Scheme: Third Year Review* (Review Report, Kimberley Process, November 2006).

Ad Hoc Working Group on the Review of the Kimberley Process Certification Scheme, *Kimberley Process Certification Scheme Questionnaire for the Review of the Scheme* (Questions for Review Submissions, Kimberley Process, 2005).

Kimberley Process, *Administrative Decision: Implementation of Peer Review in the Kimberley Process* (Plenary Meeting Decision, Sun City, South Africa, 30 October 2003).

Kimberley Process, *Administrative Decision: Participation Committee Terms of Reference* (Plenary Meeting Decision, Gatineau, Quebec, 29 October 2004).

Kimberley Process, *Administrative Decision: Terms of Reference Ad-Hoc Working Group on the Review of the KPCS* (Revised 31 July 2006).

Kimberley Process, *Administrative Decision about the KPCS Peer Review System* (Plenary Meeting Decision, Gaborone, Botswana, November 2006).

Kimberley Process, *Administrative Decision on Statistical Reporting* (undated).

Kimberley Process, *A Note from Ambassador Milovanovic* (30 July 2013). Available at: www.kimberleyprocess.com/en/note-ambassador-milovanovic.

Kimberley Process, *Diamond Store Guarantees Ethical, Conflict-free Canadian Diamonds* (Press Release, undated). Available at: www.kimberleyprocess.com/en/system/files/documents/Diamond%20Store%20guarantees%20ethical,%20conflict-free%20diamonds.pdf.

Kimberley Process, *Interlaken Declaration of 5 November 2002 on the Kimberley Process Certification Scheme for Rough Diamonds* (5 November 2002).

Kimberley Process, *The Kimberley Process Certification Scheme* (Core Document, 2002).

Kimberley Process, *The Kimberley Process Certification Scheme: Third Year Review,* Kimberley Process (November 2006) 17. Available at: www.state.gov/documents/organization/77156.pdf.

Kimberley Process, *Rules of Procedure of Meetings of the Plenary, and its Ad Hoc Working Groups and Subsidiary Bodies* (2003).

Kimberley Process, *Working Document no 3/2001* (21 August 2001).

Kimberley Process, *Working Group on Statistics Terms of Reference* (29 April 2003).

Kimberley Process Secretariat, *Annual Global Summary: 2009 Production, Imports, Exports and KPC Counts* (Annual Report Summary, 7 August 2010).

Kimberley Process Working Group on Monitoring, *Submission for the 2006 Review of the KPCS* (Kimberley Process Secretariat, February 2006).

Kimberley Process Secretariat, *Annual Global Summary: 2008 Production, Imports, Exports and KPC Counts* (Annual Report Summary, 7 August 2010).

Wexler, Pamela, *The Kimberley Process Certification Scheme on the Occasion of its Third Anniversary: An Independent Commissioned Review* (Review Report submitted to the Ad Hoc Working Group on the Review of the Kimberley Process, February 2006).

Kimberley Process: Plenary and Intersessional Meetings

Kimberley Process, '2007 Kimberley Process Communique' (Brussels, Belgium, 8 November 2007).

Kimberley Process, 'Final Communique: Kimberley Process Plenary Meeting' (Sun City, South Africa, 29–31 October 2003).

Kimberley Process, 'Final Communique: Kimberley Process Plenary Meeting' (Gatineau, Canada, 29 October 2004).

Kimberley Process, 'Kimberley Process: Final Communique' (Report on Preliminary Meeting, Moscow, Russia, 3–4 July 2001).

Kimberley Process, 'Kimberley Process: Meeting in Twickenham' (Report on Preliminary Meeting, Twickenham, England, 11–13 September 2001).

Kimberley Process, 'Kimberley Process Communique' (New Delhi, India, 6 November 2008).

Kimberley Process Secretariat, 'Kimberley Process Intersessional Meeting Communique' (Windhoek, Namibia, 25 June 2009).

Kimberley Process, 'Kimberley Process Meeting: Final Communique' (Report on Preliminary Meeting, Brussels, Belgium, 25–27 April 2001).

Kimberley Process, 'Kimberley Process Meeting: Final Communique' (Report on Preliminary Meeting, Ottawa, Canada, 18–20 March 2002).

Kimberley Process, 'Kimberley Process Meeting and Technical Workshop' (Report on Preliminary Meeting, Windhoek, Namibia, 13–16 February 2001).

Kimberley Process, 'Kimberley Process Plenary: Final Communique' (Gaborone, Botswana, 6–9 November 2006).

Kimberley Process, 'Kimberley Process Plenary: Session Communique' (Swakopmund, Namibia, 5 November 2009).

Kimberley Process, 'Kimberley Process Plenary Meeting: Final Communique', (Moscow, 2005). Available at: www.kimberleyprocess.com/en/2005-final-communique-plenary-moscowcompendium.

Kimberley Process Chair, 'Report on the Intersessional Meeting of the Working Groups of the Kimberley Process Certification Scheme' (New Delhi, India, 19 June 2008).

Kovanda, Karel, Kimberley Process Chair, 'Valedictory Remarks of Mr Karel Kovanda, Kimberley Process 2007 Chairman' (Intersessional Meeting, Brussels, Belgium, 8 November 2007).

Kimberley Process: Notices from the Chair

Bernhardt Esau MP, Kimberley Process Chair, 'Clarification About the Kimberley Process Chair's Working Visit to Zimbabwe Which Took Place 19–21 August 2009' (Kimberley Process Secretariat, Windhoek, Namibia, 3 September 2009).

Bernhardt Esau MP, Kimberley Process Chair, 'Public Statement on the Situation in the Marange Diamond Fields, Zimbabwe' (Kimberley Process Secretariat, Windhoek, Namibia, 26 March 2009).

Bernhardt Esau MP, Kimberley Process Chair, 'Risk of Fake KP Certificates' (Kimberley Process Secretariat, Windhoek, Namibia, June 2009).

Bernhardt Esau MP, Kimberley Process Chair, 'Statement: High Level Envoy Visit to Zimbabwe: Situation in Marange Diamond Fields' (Kimberley Process Secretariat, Windhoek, Namibia, 16 April 2009).

Boaz Hirsch, Kimberley Process Chair, 'Appointment of KP Monitor to Zimbabwe' (Kimberley Process Secretariat, Jerusalem, Israel, 1 March 2010).

Boaz Hirsch, Kimberley Process Chair, 'Re: Trade of Marange Diamonds in Compliance with KPCS Requirements: Vigilance Against the Laundering of Illicit Shipments' (Letter to Kimberley Process Participants, 6 May 2010).

Karel Kovanda, Kimberley Process Chair, 'The Appearance of Fraudulent Certificates' (Letter to Kimberley Process Members, 23 March 2007).

Khullar Rahul, Final Message from Kimberley Process Chair to Kimberley Process Members, 31 December 2008.

Khullar Rahul, Message Regarding Compliance of Venezuela from Kimberley Process Chair to Kimberley Process Members, 9 July 2008.

Letter Prior to Plenary Meeting in Brussels from Karel Kovanda, Kimberley Process Chair to Kimberley Process Members, 2007.

Letter from Karel Kovanda, Kimberley Process Chair, to Kimberley Process Members, 31 January 2007.

Letter from Karel Kovanda, Kimberley Process Chair, to Kimberley Process Members, 1 March 2007.

Domestic Legislation

Code Minier 1995 [Mining Code] (Côte d'Ivoire) Law No 95-553, 17 July 1995.

Da Lei dos Diamantes 1994 [The Diamond Law] (Angola) Law No 1/92, 7 October 1994.

Lei das Actividades Geológicas e Mineiras 1992 [Law on Geological and Mining Activities] (Angola) Law No 1/92, 17 January 1992.

Lei sobre o Regime Especial das Zonas de Reserva Diamantífera 1994 [Law on the Special Regime of Zones of Diamondiferous Deposits] (Angola) Law No 17/94, 7 October 1994.

Loi Portant Code Minier 2002 [Law Relating to the Mining Code] (The Democratic Republic of Congo) Law No 007/2002, 11 July 2002.

Mines and Minerals Act 1994 (Sierra, Leone).

Regime Aduaneiro Aplicável ao Sector Mineiro 1996 [Customs Regime Applicable to the Mining Sector] (Angola) Decree-Law No 12-B/96, 31 May 1996.

Regulamento do Regime Fiscal para a Indústria Mineira 1996 [Regulation of the Fiscal Regime for the Mining Industry] (Angola) Decree-Law No 4-B/96, 31 May 1996.

Treaties

Geneva Convention Relative to the Treatment of Prisoners of War (Geneva Convention III), opened for signature 12 August 1949, 75 UNTS 135 (entered into force 21 October 1950).

Hague Convention (IV) Respecting the Laws and Customs of War on Land and Its Annex: Regulations Concerning the Laws and Customs of War on Land, opened for signature 18 October 1907 (entered into force 26 January 1910).

Protocol Additional to the Geneva Conventions of 12 August 1949, and relating to the Protection of Victims of Non-International Armed Conflicts (Protocol II), opened for signature 8 June 1977, 1125 UNTS 609 (entered into force 7 December 1978).

Rome Statute of the International Criminal Court, opened for signature 17 July 1998, 2187 UNTS 90 (entered into force 1 July 2002).

Vienna Convention on the Law of Treaties, opened for signature 23 May 1969, 1155 UNTS 331 (entered into force 27 January 1980).

Non-Governmental Organisation, Industry, Government and Media Reports

Africa Initiative Programme and Forum on Early Warning and Early Response (FEWER-Africa), 'Elections and Security in Ituri: Stumbling Blocks and Opportunities for Peace in the Democratic Republic of Congo' (Report, 13 June 2006).

Allen, M, 'The "Blood Diamond" Resurfaces', *Wall Street Journal*, (18 June 2010). Available at: www.wsj.com/articles/SB1000142405 2748704198004575311282588959188.

Australian Law Reform Commission, 'Penalties: Policy, Principles and Practice in Government Regulation' (Conference Discussion Paper, June 2001).

Black, J, 'Critical Reflections on Regulation', *Australian Journal of Legal Philosophy,* vol. 27, 2002, 20.

Braithwaite, John, 'Realism and Principled Engagement in International Affairs and the Social Sciences of Regulation' (Paper presented at RegNet@10 Conference, The Australian National University, March 2011).

Brilliant Earth, *Conflict Diamond Issues* (2007). Available at: www. brilliantearth.com/conflict-diamond-child-labor/.

British Broadcasting Corporation, 'Eastern Congo Peace Deal Signed' (23 January 2008). Available at: www.globalpolicy.org/security/ issues/congo/2008/0123gomadeal1.htm.

British Broadcasting Corporation, 'Gbagbo Held After Assault on Residence' (11 April 2011). Available at: www.bbc.com/news/ world-africa-13039825.

British Broadcasting Corporation, 'Profile: Guus van Kouwenhoven' (2008). Available at: news.bbc.co.uk/2/hi/africa/5055442.stm.

British Broadcasting Corporation, 'Q&A: DR Congo Conflict', *BBC News: World Edition* (15 December 2004). Available at: news. bbc.co.uk/2/hi/africa/3075537.stm.

British Broadcasting Corporation, 'Q&A: Ivory Coast's Crisis', *BBC News: World Edition* (17 January 2006). Available at: news.bbc.co.uk/2/hi/africa/3567349.stm.

British Broadcasting Corporation, 'Q&A: Plunder in the Congo', *BBC News: World Edition* (21 October 2002). Available at: news.bbc.co.uk/2/hi/business/2346817.stm.

C, Jessica, *Côte D'Ivoire Diamonds Remain Conflict Diamonds* (2 May 2011) Diamond Price Guide: A World of Diamond Experts. Available at: www.diamondpriceguide.com/news/nc43_Precious-Metals/n89857_Cote-DIvoi.

Charbonneau, Louis, 'Zimbabwe "Blood Diamonds" Dispute Breaks out at UN', *Reuters Canada* (11 December 2009). Available at: ca.reuters.com/article/topNews/idCATRE5BA3OI20091211.

Davies, Catherine, *Reciprocal Violence: Mass Expulsions Between Angola and DRC* (17 February 2011) The Human Rights Brief. Available at: hrbrief.org/2011/02/reciprocal-violence-mass-expulsions-between-angola-and-the-drc/.

Dawn News, 'Blood Diamond Fears in Ivory Coast Political Duel' (28 December 2010). Available at: www.dawn.com/2010/12/28/blood-diamond-fears-in-ivory-coast-political-duel.html.

De Beers Diamond Jewellers Ltd, 'Annual Report' (1992).

De Beers Diamond Jewellers Ltd, 'Written Testimony before the United States Congress, House Committee on International Relations, Subcommittee on Africa', *Hearing into the Issue of 'Conflict Diamonds'* (25 May 2000). Available at: www.diamonds.net/fairtrade/Article.aspx?ArticleID=4046.

Diamond Development Initiative International, *Accra Conference Background Note,* (October 2005). Available at: www.artisanalmining.org/userfiles/file/DDI_Accra_Oct.5.pdf.

Diamond Development Initiative International, *2008 Annual Report: Beyond Dreams the Journey Begins* (2008).

EITI, *EITI Countries* (2009). Available at: eiti.org/countries.

Fonseca, A, *Four Million Dead: The Second Congolese War, 1998-2004* (18 April 2004) 49. Available at: www.oocities.org/afonseca/Congo War.htm.

Getty Images, *Charles Taylor: Photos* (4 June 2007). Available at: www. daylife.com/photo/08ECafbd1Mdh8?q=Charles+Taylor.

Global Policy Forum, *The Democratic Republic of Congo* (2006). Available at: www.globalpolicy.org/security/issues/kongidx.htm.

Global Witness, 'Broken Vows: Exposing the "Loupe Holes" in the Diamond Industry's Efforts to Prevent the Trade in Conflict Diamonds' (Report, March 2004).

Global Witness, 'Cautiously Optimistic: The Case for Maintaining Sanctions in Liberia' (Report, 2006).

Global Witness, *Conflict Diamond Scheme Must Resolve Zimbabwe Impasse* (5 November 2010). Available at: www.globalwitness. org/library/conflict-diamond-scheme-must-resolve-zimbabwe-impasse.

Global Witness, 'Conflict Diamonds: Possibilities for the Identification, Certification and Control of Diamonds' (Report, June 2000).

Global Witness, 'For a Few Dollars More: How Al Qaeda Moved into the Diamond Trade' (Report, April 2003).

Global Witness, 'Global Witness Leaves Kimberley Process, Calls for Diamond Trade to be Held Accountable' (2 December 2011). Available at: https://www.globalwitness.org/archive/global-witness-leaves-kimberley-process-calls-diamond-trade-be-held-accountable/.

Global Witness, *Global Witness Welcomes Dutch Court's Decision to Hear New Prosecution witnesses in Kouwenhoven Case*. Available at: www. globalwitness.org/en/archive/global-witness-welcomes-dutch-courts-decision-hear-new-prosecution-witnesses-kouwenhoven/.

Global Witness, *Global Witness Welcomes French 'Angolagate' Verdict as Victory for Justice* (28 October 2009). Available at: www. globalwitness.org/en/archive/global-witness-welcomes-french-angolagate-verdict-victory-justice/.

Global Witness, *Industry Must Refuse Zimbabwe Diamonds Certified by Rogue Monitor* (16 November 2010). Available at: www.globalwitness.org/library/industry-must-refuse-zimbabwe-diamonds-certified-rogue-monitor.

Global Witness, 'Kimberley Process Certification Scheme Questionnaire for the Review of the Scheme' (Review Submission, 5 April 2006).

Global Witness, *Natural Resources and Conflict under the Legal Spotlight, War Crimes Trial of Gus Kouwenhoven to Commence in The Hague* (21 April 2006). Available at: www.globalwitness.org/library/natural-resources-and-conflict-under-legal-spotlight-war-crimes-trial-gus-kouwenhoven.

Global Witness, 'Return of the Blood Diamond: The Deadly Race to Control Zimbabwe's New-found Diamond Wealth' (Report, 14 June 2010).

Global Witness, 'A Rough Trade: The Role of Companies and Governments in the Angolan Conflict' (Report, Global Witness Ltd, 1998).

Global Witness and Amnesty International, 'Déjà vu: Diamond Industry Still Failing to Deliver on Promises: Summary of UK and US Results of UK and US Results of Global Witness and Amnesty International Survey' (Report, Global Witness, October 2004).

Global Witness and Partnership Africa Canada, 'The Key to the Kimberley: Internal Diamond Controls: Seven Case Studies' (Report, 2004).

Global Witness and Partnership Africa Canada, 'Rich Man, Poor Man, Development Diamonds and Poverty Diamonds: The Potential for Change in the Artisanal Alluvial Diamond Fields of Africa' (Report, October 2004).

Hennessy, Selah, *Congo's Diamond Industry Let Back into Kimberley Process* (9 November 2007) Global Policy Forum. Available at: www.globalpolicy.org/security/issues/congo/2007/1109drckp.htm.

Herskovitz, Jon, 'Zimbabwe Govt, c.bank in Blood Diamond Trade: WikiLeaks', *Reuters: Africa* (9 December 2010). Available at: af.reuters.com/article/topNews/idAFJOE6B80F820101209.

Human Rights Watch, 'Diamonds in the Rough: Human Rights Abuses in the Marange Diamond Fields of Zimbabwe' (Report, June 2009).

Hussain, Sakina Sadat, 'A Diamond's Journey: Grime Behind the Glitter?', *World News* (26 June 2009). Available at: www.msnbc.msn.com/id/15842528.

International Committee of the Red Cross, *Rule 1. The Principle of Distinction between Civilians and Combatants.* Available at: www.icrc.org/customary-ihl/eng/docs/v1_ru_rule1.

McClanahan, Paige, *As Ivory Coast's Gbagbo Holds Firm, 'Blood Diamonds' Flow for Export* (23 January 2011) ReliefWeb. Available at: reliefweb.int/node/381665.

Montague, D and F Berrigan, 'The Business of War in the Democratic Republic of Congo', *Dollars and Sense Magazine* (July/August 2001). Available at: www.thirdworldtraveler.com/Africa/Business_War_Congo.html.

Partnership Africa Canada, *Diamond Development Initiative International.* Available at: www.pacweb.org.

Partnership Africa Canada, 'Diamond Industry Annual Review: Republic of Angola 2005' (Annual Report, July 2005).

Partnership Africa Canada, 'Diamonds and Clubs: The Militarised Control of Diamonds and Power in Zimbabwe' (Report, June 2010).

Partnership Africa Canada, 'Diamonds and Human Security: Annual Review 2009' (Annual Report, 2009).

Partnership Africa Canada, 'Diamonds and Human Security: Annual Review 2008' (Annual Report, October 2008).

Partnership Africa Canada, 'Kimberley Process lets Zimbabwe off the Hook Again', (2 November 2011). Available at: www.pacweb.org/Documents/Press_releases/2011/KP_lets_Zim_off_the_hook_Nov2011.pdf.

Partnership Africa Canada, *Other Facets*, Ottawa (October 2006).

Partnership Africa Canada, *Other Facets: News and Views on the International Effort to End Conflict Diamonds*, No. 34 (February 2011).

Partnership Africa Canada, 'PAC and the Kimberley Process: A History'. Available at: www.pacweb.org/en/pac-and-the-kimberly-process.

Partnership Africa Canada, 'Zimbabwe, Diamonds and the Wrong Side of History' (Report, March 2009).

Partnership Africa Canada and CENADEP, 'Diamond Industry Annual Review: Democratic Republic of Congo 2007' (Annual Report, September 2007).

Ray, Nandita, *The Kimberley Process Certification* (2004) JewelInfo4U. Available at: www.jewelinfo4u.com/The_Kimberley_Process_Certification.aspx.

Scott, Colin, 'Regulating in Global Regimes', Working Paper No 25/2010, University College Dublin (2010).

Simpson, John, 'Profiting from Zimbabwe's "blood diamonds"', *BBC News* (online), 20 April 2009. Available at: news.bbc.co.uk/2/hi/africa/8007406.stm.

Smillie, Ian, 'Paddles for Kimberley: An Agenda for Reform' (Report, Partnership Africa Canada, June 2010).

Smillie, I, L Gberie and R Hazleton, 'The Heart of the Matter: Sierra Leone, Diamonds and Human Security' (Report, Partnership Africa Canada, 2000).

Stempel, J, 'Russian Arms Dealer Viktor Bout's U.S. Conviction Upheld', Reuters, 27 September 2013. Available at: www.reuters.com/article/us-usa-crime-bout-idUSBRE98Q0PG20130927.

Taylor, Telford, 'Final Report to the Secretary of the Army on the Nuernberg War Crimes Trials Under Control Council Law No. 10' (William S. Hein & Co: Buffalo, 1997).

Wikipedia, *Jungfrau* (30 August 2008). Available at: en.wikipedia.org/wiki/File:2008_Jungfrau.jpg.

Responses to Research Questions

Interview with Global Witness Representative (telephone interview, 21 May 2007).

Interview with Global Witness Representative (telephone interview, 30 April 2007).

Interview with Rio Tinto Representative (telephone interview, 30 May 2007).

Written Response from Australian Government to Author's Interview Questions (14 September 2007).

United Nations General Assembly Resolutions

Report of the Kimberley Process Certification Scheme to the General Assembly pursuant to resolution 61/28, by Fernando M. Valenzuela on behalf of Kimberley Process Chair, 62nd sess, Agenda Item 13, UN Doc A/62/543 (13 November 2007).

The Role of Diamonds in Fuelling Conflict: Breaking the Link Between the Illicit Transaction of Rough Diamonds and Armed Conflict as a Contribution to Prevention and Settlement of Conflicts, GA Res 55/56, UN GAOR, 55th sess, 79th plen mtg, Un Doc A/RES/55/56 (1 December 2001).

The Role of Diamonds in Fuelling Conflict: Breaking the Link Between the Illicit Transaction of Rough Diamonds and Armed Conflict as a Contribution to Prevention and Settlement of Conflicts, GA Res 56/263, UN GAOR, 56th sess, 96th plen mtg, UN Doc A/Res/56/263 (13 March 2002).

The Role of Diamonds in Fuelling Conflict: Breaking the Link Between the Illicit Transaction of Rough Diamonds and Armed Conflict as a Contribution to Prevention and Settlement of Conflicts, GA Res 57/302, UN GAOR, 57th sess, 83rd plen mtg, UN Doc A/RES/57/302 (15 April 2003).

The Role of Diamonds in Fuelling Conflict: Breaking the Link Between the Illicit Transaction of Rough Diamonds and Armed Conflict as a Contribution to Prevention and Settlement of Conflicts, GA Res 58/290, UN GAOR, 58th sess, 85th plen mtg, UN Doc A/RES/58/290 (14 April 2004).

The Role of Diamonds in Fuelling Conflict: Breaking the Link Between the Illicit Transaction of Rough Diamonds and Armed Conflict as a Contribution to Prevention and Settlement of Conflicts, GA Res 60/182, UN GAOR, 60th sess, 67th plen mtg, UN Doc A/RES/60/182 (20 December 2005).

The Role of Diamonds in Fuelling Conflict: Breaking the Link Between the Illicit Transaction of Rough Diamonds and Armed Conflict as a Contribution to Prevention and Settlement of Conflicts, GA Res 61/28, UN GAOR, 61st sess, 64th plen mtg, UN Doc A/RES/61/28 (4 December 2006).

The Role of Diamonds in Fuelling Conflict: Breaking the Link Between the Illicit Transaction of Rough Diamonds and Armed Conflict as a Contribution to Prevention and Settlement of Conflicts, 62nd sess, Agenda Item 13, UN Doc A/62/L.16 (21 November 2007).

The Role of Diamonds in Fuelling Conflict: Breaking the Link Between the Illicit Transaction of Rough Diamonds and Armed Conflict as a Contribution to Prevention and Settlement of Conflicts, 63rd sess, Agenda Item 11, UN Doc A/63/L.52 (5 December 2008).

United Nations Security Council Resolutions and Reports

SC Res 827, UN SCOR, 3217th mtg, UN Doc S/RES/827 (25 May 1993) ('Statute of the International Criminal Tribunal for the Former Yugoslavia').

SC Res 864, UN SCOR, 3277th mtg, UN Doc S/RES/864 (5 June 1998) (Angola).

SC Res 1127, 3814th mtg, UN Doc S/RES/1127 (28 August 1997) (Angola).

SC Res 1171, UN SCOR, 3889th mtg, UN Doc S/RES/1171 (5 June 1998) (Sierra Leone).

SC Res 1173, UN SCOR, 3891st mtg, UN Doc S/RES/1173 (12 June 1998) (Angola).

SC Res 1176, UN SCOR, 3894th mtg, UN Doc S/RES/1176 (24 June 1998).

SC Res 1295, UN SCOR, 4129th mtg, UN Doc S/RES/1295 (18 April 2000) (Angola).

SC Res 1306, UN SCOR, 4168th mtg, UN Doc S/RES/1306 (5 July 2000) (Sierra Leone).

SC Res 1343, UN SCOR, 4287th mtg, UN Doc S/RES/1343 (7 March, 2001) (Liberia).

SC Res 1385, UN SCOR, 4442nd mtg, UN Doc S/RES/1385 (19 December 2001) (Sierra Leone).

SC Res 1408, UN SCOR, 4256th mtg, UN Doc S/RES/1408 (6 May 2002) (Liberia).

SC Res 1457, UN SCOR, 4691st mtg, UN Doc S/RES/1457 (24 January 2003) (Democratic Republic of Congo).

SC Res 1459, UN SCOR, 6494th mtg, UN Doc S/RES/1459 (28 January 2003) (Kimberley Process Certification Scheme).

SC Res 1521, UN SCOR, 4890th mtg, UN Doc S/RES/1521 (22 December 2003) (Liberia).

SC Res 1579, UN SCOR, 5105th mtg, UN Doc S/RES/1579 (21 December 2004) (Liberia).

SC Res 1643, UN SCOR, 5327th mtg, UN Doc S/RES/1643 (15 December 2005) (Côte d'Ivoire).

Final Report of the Group of Experts on the Democratic Republic of the Congo, UN Doc S/2008/773 (21 November 2008).

Final Report Panel of Experts on the Illegal Exploitation of Natural Resources and Other Forms of Wealth of the Democratic Republic of the Congo, UN Doc S/2002/1146 (16 October 2002).

Interim Report of the Group of Experts on the Democratic Republic of the Congo, Pursuant to Security Council Resolution 1698 (2006), UN Doc S/2007/40 (31 January 2007).

Lusaka Protocol, UN SCOR, 17th sess, UN Doc S/1994/1441 (22 December 1994).

Progress Report of the Secretary-General on the United Nations Observer Mission in Angola (MONUA), UN Doc S/1997/640 (13 August 1997).

Report of the Group of Experts on Côte d'Ivoire pursuant to Paragraph 11 of Security Council Resolution 1946 (2010), UN Doc S/2011/272 (17 March 2011).

Report of the Group of Experts Submitted Pursuant to Paragraph 7 of Security Council Resolution 1584 (2005) Concerning Côte d'Ivoire, UN Doc S/2005/699 (7 November 2005).

Report of the Panel of Experts Appointed Pursuant to Security Council Resolution 1306 (2000), paragraph 19, in Relation to Sierra Leone, UN Doc S/2000/1195 (20 December 2000).

Report of the Panel of Experts on the Illegal Exploitation of Natural Resources and Other Forms of Wealth of the Democratic Republic of Congo, UN Doc S/2001/357 (12 April 2001).

Report of the Panel of Experts on Violations of Security Council Sanctions Against UNITA, UN Doc S/2000/203 (10 March 2000).

Supplementary Report of the Monitoring Mechanism on Sanctions Against UNITA, UN Doc 2001/966 (12 October 2001).

Update Report of the Group of Experts Submitted Pursuant to Paragraph 2 of Security Council Resolution 1632 (2005) Concerning Côte d'Ivoire, UN Doc S/2006/204 (31 March 2006).

The Role of Diamonds in Fuelling Conflict: Statement to General Assembly, Nirupam Sen, Kimberley Process Chair, 63rd sess, 67th plen mtg, Agenda Item 11, UN Doc A/63/PV.67 (11 December 2008).

United Nations Office of the High Commissioner for Human Rights, Report of the Mapping Exercise documenting the Most Serious Violations of Human Rights and International Humanitarian Law Committed within the Territory of the Democratic Republic of the Congo between March 1993 and June 2003 (August 2010).

www.ingramcontent.com/pod-product-compliance
Lightning Source LLC
Chambersburg PA
CBHW040149270326
41928CB00032B/3295